Kettlebell Strength Training

ANATOMY

Michael Hartle, DC

HUMAN KINETICS

Library of Congress Cataloging-in-Publication Data

Names: Hartle, Michael, 1967- author.
Title: Kettlebell strength training anatomy / Michael Hartle.
Description: Champaign, IL : Human Kinetics, [2024]
Identifiers: LCCN 2023004343 (print) | LCCN 2023004344 (ebook) | ISBN 9781718208599 (paperback) | ISBN 9781718208605 (epub) | ISBN 9781718208612 (pdf)
Subjects: LCSH: Kettlebells. | Weight training.
Classification: LCC GV547.5 .H37 2024 (print) | LCC GV547.5 (ebook) | DDC 613.713--dc23/eng/20230209
LC record available at https://lccn.loc.gov/2023004343
LC ebook record available at https://lccn.loc.gov/2023004344

ISBN: 978-1-7182-0859-9 (print)

Senior Acquisitions Editor: Michelle Earle; **Senior Developmental Editor:** Cynthia McEntire; **Managing Editor:** Shawn Donnelly; **Copyeditor:** Janet Kiefer; **Permissions Manager:** Laurel Mitchell; **Graphic Designer:** Julie L. Denzer; **Cover Designer:** Keri Evans; **Cover Design Specialist:** Susan Rothermel Allen; **Photograph (for cover illustration reference):** Human Kinetics/Illustrator: Heidi Richter; **Photo Production Manager:** Jason Allen; **Senior Art Manager:** Kelly Hendren; **Illustrations:** © Human Kinetics/Heidi Richter; **Printer:** Versa Press

Human Kinetics books are available at special discounts for bulk purchase. Special editions or book excerpts can also be created to specification. For details, contact the Special Sales Manager at Human Kinetics.

Printed in the United States of America 10 9 8 7 6 5 4 3 2 1

The paper in this book is certified under a sustainable forestry program.

Human Kinetics
1607 N. Market Street
Champaign, IL 61820
USA

United States and International
Website: **US.HumanKinetics.com**
Email: info@hkusa.com
Phone: 1-800-747-4457

Canada
Website: **Canada.HumanKinetics.com**
Email: info@hkcanada.com

E8412

Kettlebell Strength Training

ANATOMY

CONTENTS

Foreword vi

Preface viii

Acknowledgments x

Introduction xi

CHAPTER 1 TRAINING WITH KETTLEBELLS 1

CHAPTER 2 DEADLIFT 13

CHAPTER 3 SWING 35

CHAPTER 4 CLEAN AND PRESS 59

CHAPTER 5 GET-UP 95

CHAPTER 6 **SQUAT** 115

CHAPTER 7 **SNATCH** 135

CHAPTER 8 **ROW AND PULL-UP** 153

CHAPTER 9 **CARRY** 165

CHAPTER 10 **MOBILITY** 181

Exercise Finder 197

About the Author 199

Earn Continuing Education Credits/Units 201

FOREWORD

More than a decade ago, we developed a radically different kind of plank exercise at our School of Strength. When Dr. Bret Contreras compared the electromyography in the traditional plank and our hard-style plank, he discovered that in the latter, the internal obliques fired twice as strong, the abdominal muscles three times as strong, and the external obliques four times as strong as in the traditional plank.

Then our master instructor, Dr. Michael Hartle, added a finishing touch. One more cue—and the already off-the-chart recruitment jumped up another notch . . .

The point of this story is this: If you are looking for an expert who can translate theoretical knowledge of the neuromuscular system into very practical strength, Dr. Michael Hartle is your man, and *Kettlebell Strength Training Anatomy* is your book.

I have had the pleasure of knowing Doc Hartle for a decade and a half as a colleague and a friend. His rare combination of knowledge, experience, and innovation make him stand out among strength and conditioning professionals.

My first reason for feeling this way is that he is a doctor of chiropractic (DC). When young Americans aspiring to become strength coaches ask me what sort of education they should pursue, I name a chiropractic degree as one of their top choices. Over the years, I have met a number of DCs who were also strength athletes. They were unfailingly successful on the platform or the field due to their exquisite knowledge of the neuromuscular system. For many other health professionals, this knowledge is rather theoretical; chiropractors literally feel it on their fingertips. And when they choose to direct it to an adjacent field of strength and conditioning, records fall. This brings me to my second reason . . .

Dr. Hartle is a national bench press champion with the best lift of 535 pounds. He also won silver in a cutthroat three-lift national powerlifting competition. As if that was not enough of a challenge, Doc played semiprofessional football—until he was older than most other players' fathers.

My third reason for admiring his credentials is that Doc Hartle, not content being an elite athlete himself, has strength coached many others to excellence: lifters, football players, track and field competitors, and other athletes. He was head coach of the USA Powerlifting national bench press team at eight consecutive World Bench Press Championships, a rare distinction indeed.

Not surprisingly, such a sharp mind always stays on the cutting edge of training. This is why Doc joined our kettlebell School of Strength long before the kettlebell became a household name—when most of the readers of *Kettlebell Strength Training Anatomy* were too young to drive. In short order, Doc joined the elite group of our master instructors and later helped refine our curricula. He coauthored the exercise technique section of the *StrongFirst Lifter* instructor manual, the most precise and detail-intensive manual in the industry.

Ladies and gentlemen, I am honored to present this fine volume, *Kettlebell Strength Training Anatomy*, by Dr. Michael Hartle, StrongFirst Certified Master Instructor, bench press national champion, and Team USA head coach.

Pavel Tsatsouline, CEO, StrongFirst, Inc.
Author of *Kettlebell Simple & Sinister*

PREFACE

As I was reading a powerlifting magazine in 2003, I came across an article written by Pavel Tsatsouline regarding using kettlebells to enhance one's training for the sport of powerlifting. In the article, he discussed using these cannonball-shaped objects with a handle to better my mobility for the back squat, using kettlebell presses and get-ups to improve my bench press, and using swings, including heavy swings, to increase my deadlift ability.

At first, I was skeptical: How could these small pieces of iron help improve my lifting ability? I say *small* because when you are used to lifting hundreds of pounds, it's hard to believe that this cannonballed-shaped object that usually weighs between 16 and 48 kilograms (35-106 lb) could possibly make your powerlifting movements stronger! I put the magazine down and didn't think about kettlebells again until 2004, when I attended the NSCA national conference, where Pavel's kettlebell company had a booth in the exhibition area. I met Pavel and purchased two 24-kilogram kettlebells and then placed them in my gym at my clinic. Honestly, they sat there and gathered dust for a year and a half before I decided to take Pavel's invitation to attend his kettlebell certification course.

In April of 2006, I went to St. Paul, Minnesota, and attended the certification course. We were using 24-, 32-, and 40-kilogram kettlebells throughout the three-day weekend. Initially I was not impressed with the weight being used since I was a national-level powerlifter and was used to lifting hundreds of pounds. Pavel certainly changed my initial impression. I was accustomed to lifting with symmetry, meaning my hands were equidistant from each other on the barbell and my feet were equally distanced from each other on the platform. The offset nature of the kettlebell proved to me this piece of equipment was going to be another powerful tool in my arsenal of training implements.

The collection of kettlebells at my gym slowly grew. Every time I drove to Minneapolis to see my family, I would make a stop at the headquarters to purchase new kettlebells, which I would bring back to my clinic gym for my personal use and for my family and my staff to use. That summer of 2006, I decided to retire from powerlifting competition and try my hand at semipro football at the young age of 38. Even though I had decided to play football, the mental aspect of letting my absolute strength gains over the last 20 years go by the wayside so I could play three hours of American football in the summer heat was tough.

Once I wrapped my head around that decision, the integration of kettlebells with my barbell training enhanced my football playing on the field. I needed to increase my strength endurance to not only improve my performance on the field but also decrease the chances of injury. During the time I was a powerlifting athlete, the barbell had never attacked me. However, with football, I was being hit, pushed, and blocked while being a defensive tackle on Saturday nights. Then on Monday I was back at my clinic, treating patients. Using kettlebells in addition to barbells to build my strength and endurance paid off dividends: The only injury I had during my 10 years of playing football was a sprained ankle, and my performance improved every year.

Fast-forward 17 years, and kettlebells have an even larger place in my everyday life. I am now a StrongFirst Certified Master Instructor with Pavel's company, StrongFirst. Several of the members of my clinic are now certified kettlebell instructors. I train with kettlebells two to three days per week, and they are used daily in helping to rehabilitate my patients. And I teach students at various locations around the world how to become certified kettlebell instructors.

There was a recent commercial on TV in which a grandfather was seen going into his garage daily to train by picking up a kettlebell. The main reason for him doing this was so that he would have the strength to pick up and carry his granddaughter. This meant a lot to him, which is why he did it on a daily basis, and the results were eventually shown; he picked up his beloved granddaughter. The grandfather's use of his garage for his kettlebell training demonstrates one of the beautiful things about kettlebells: They are easily transported to a city park, to a friend's backyard, or to anywhere where you want to do your training. Kettlebells are made in specific weights, and most people only need two or three kettlebells to get a fantastic training session done.

The idea for this book came from Michelle Earle, a senior acquisitions editor at Human Kinetics. She had seen a chapter I wrote for Dr. Craig Liebenson's book, *Rehabilitation of the Spine*, and she wanted to know if I would be interested in writing a book about kettlebells. At first, the idea both excited me and worried me since I didn't know if I would have enough time to devote to completing the project. Michelle and her staff have been great to work with, especially when I decided to go back to school while being a full-time chiropractic physician and teaching for StrongFirst to pursue getting my PhD in exercise science while in the midst of writing this book. What you are reading is a labor of love and a passion—not only for strength training but for training with kettlebells. I hope you enjoy it and use the knowledge and experience that are shared with you throughout these pages. Have a strong day!

ACKNOWLEDGMENTS

I dedicate this book to my three sons, Colin, Anthony, and Marcus. It has been a joy raising you to become the men that you are. Thank you to my mom and dad for raising me to become the man that I am. I also want to thank Pavel Tsatsouline for not only teaching me what kettlebells were and how to use them but more about strength training than I knew previously.

Thank you to all my fellow StrongFirst instructors for not only being great students but also teaching me in return. I am always a student! Special thanks to all the Master Instructors, especially Brett Jones, my original Team Leader; Fabio Zonin, a great friend and peer; Pavel Macek, someone who is always there for me; and Karen Smith, whose heart is bigger than herself.

Thank you to Michelle Earle at Human Kinetics and her wonderful team. In the midst of writing this book, I decided to pursue my PhD in exercise science on top of everything else I had going on at that time. I greatly appreciate their patience and skills in helping move this book from just an idea to what you are holding in your hand.

Finally, thank you to my staff, Holli, Joshua, Micalah, and Katy, at my clinic for putting up with me while I was writing this book. Your help in telling me where to go and what to do during the day made it possible for me to write this at night and on the weekends. And to Amy, thank you for being understanding and giving me support while I wrote this book.

INTRODUCTION

When I was first introduced to kettlebells in 2003, I personally thought they were designed weirdly and were misshaped. Little did I know then that 20 years later I would be teaching students how to lift these cannonball-shaped objects with a handle. To say that these kettlebells have changed my life for the better is an understatement. They also have changed the lives of each individual I have had the honor of teaching, not only to become stronger but to help improve the quality of their life overall.

This book is designed for the person who has no idea what either a kettlebell is or how to use it—and also for the experienced kettlebell user who wants to learn how to do a particular exercise to maximize their performance. For the former my suggestion is to read the book cover to cover, stopping at each chapter to not only practice but gain experience in each lift. Regardless of your experience with kettlebells, I encourage you to read chapter 1 from start to finish. It will give you the base layer, or foundation, to be successful in not only understanding but performing the exercises in chapters 2 through 10. I would dog-ear the first page of chapter 1 so you can reference it as needed as you use this book.

To help you understand what you'll find in these pages, here's a summary of each chapter.

Chapter 1 discusses training with kettlebells, which not only creates a foundation for the reader but also discusses items such as hand care, the advantages of kettlebell training, what type of footwear to wear, asymmetrical loading, and breathing behind the shield. This chapter gives a general overview of what you can expect from the successive chapters and teaches you the basics to get started in the right direction.

Chapter 2 starts off kettlebell training with learning the kettlebell deadlift and its different variations. One of the reasons I like teaching the deadlift first before any other kettlebell movement is it helps teach the student how to generate tension to be able to lift an object safely.

Chapter 3 teaches probably the most famous kettlebell exercise: the kettlebell swing. If I had to choose only one kettlebell exercise to teach a student, it would be the kettlebell swing. This, of course, does not decrease the importance of the other kettlebell exercises in this book, but it does show the importance of learning how to move through tension and relaxation and back to tension in a rapid fashion. The kettlebell swing is the dynamic or ballistic version of the kettlebell deadlift. It is also necessary to learn this exercise first, as it is utilized in almost every other kettlebell exercise in this book.

Chapter 4 teaches the student to not only perform a kettlebell clean but also to do a proper kettlebell press to help train the upper body in addition to the lower body. Having a good kettlebell clean is important, as it is a means to an end. To do a good kettlebell press, you must first do a proper kettlebell clean right before it. The same goes for the kettlebell squat.

Chapter 5 introduces the kettlebell get-up. In this exercise, you go from a horizontal position to a vertical position and back to a horizontal position while holding a weight overhead with one hand. This may sound easy, but it is not, even for seasoned kettlebell users. This exercise has shown maximal contraction of all four abdominal muscles while not performing a sit-up! Plus, what it does for your shoulder complex is truly amazing. Many athletes, from mixed-martial artists to American football players to swimmers, have had their careers reinvigorated by performing the get-up.

Chapter 6 teaches you about kettlebell squats. Unlike the kettlebell swings where you will learn how to set up in the hip-hinge position, here you will be utilizing your hips and knees in a squatting fashion. This will benefit not only your lower-body muscles but also your abdominals, which will be on fire, as their main goal is to keep you upright during the squat motion.

Chapter 7 teaches you how to perform the kettlebell snatch, the tsar of kettlebell movements. All the kettlebell exercises that have been taught so far lead up to this moment. When performed with proper technique, the kettlebell snatch will further expand your repertoire of exercises to enhance your strength and conditioning. In Chapter 1, excess postexercise oxygen consumption (EPOC) is explained in greater detail, but let's just say this: After doing several sets of kettlebell snatches, your EPOC will be greatly enhanced for a period of time. Additionally, you could also potentially be able to jump higher as a possible result of doing snatches. However, instead of me telling you, find out for yourself.

Chapter 8 goes into rows and pull-ups, and you will learn one of my favorite rowing exercises, the renegade row. From the minute I was shown this exercise in 2005 to this day, I am still perfecting my technique. Chapter 8 finishes with performing one of the best body-weight exercises there is, the pull-up. Because this is a book about kettlebell exercises, these pull-ups will be performed with a kettlebell.

The final two chapters help round out this book about kettlebell strength training anatomy. Chapter 9 deals with four different versions of kettlebell carries and their increasing level of difficulty from the beginning to the end of the chapter. While these may seem like simple exercises, don't underestimate what they can do for your body and mind.

Chapter 10 discusses various kettlebell exercises for mobility. These exercises will help you increase your mobility not only for your training session on a particular day but also for your sport or even your work performance. With 20 years of national-level powerlifting experience, followed by 10 years of playing semipro football, I can personally vouch for the five mobility exercises in this chapter. They have helped my body rehabilitate through various injuries that occurred during the 30 years of competitive athletics, and they also have kept my mobility at a level where even a 20-year-old would be jealous.

One unique feature of *Kettlebell Strength Training Anatomy* we hope you find helpful is the anatomical drawings that accompany the description of each exercise. They illustrate key muscles used during each movement, using color to highlight the engagement of primary and secondary muscle groups from start to finish (see below).

This book can be used in many different ways. A strength coach could utilize it to learn how to perform kettlebell swings and get-ups and then teach these moves to their high school athletes as they prepare for their respective sports. A grandfather could utilize it to help instruct him on how to properly pick up a kettlebell so he could become stronger to pick up his granddaughter. A new mom can utilize kettlebells to not only help her recover from the birth process but also to make her body more resilient so she can take care of her new child in a better way.

This book was written for everyone. Anyone can generate beneficial gains from using kettlebells as their chosen exercise tool, or as a supplement to barbell or body-weight exercises.

I hope you enjoy reading this book as much as I enjoyed writing it. Have a strong day!

1

TRAINING WITH KETTLEBELLS

Welcome to this book on kettlebell training. Before you get started, it is strongly suggested that you read this chapter once or twice before picking up a kettlebell and then a few more times while training. This chapter will not only go over the anatomy of the kettlebell itself, but it will also discuss hand care, the advantages of kettlebell training, and even proper technique. The proper technique part of this chapter starts with safety considerations. Please underline and highlight this section. Not only is proper technique a key to getting strong and becoming healthier, but it helps you train in a safe and effective way. In this section we will discuss breathing while lifting with the kettlebell. You will learn to breathe behind the shield and understand biomechanical breathing match. The section on asymmetrical loading is one of my favorites, especially when training athletes, as there are many sports where, at any given moment, the loading on the body is asymmetrical. When you are done reading this chapter, dog-ear the first page, because you will be referencing it time and time again.

WHY KETTLEBELL TRAINING?

Kettlebell training is a very effective and efficient exercise method with many benefits:

- You can improve your cardiovascular fitness while burning calories.
- You can complete a full training session in a relatively short amount of time.
- There is a great variety of exercises that you can perform with kettlebells.
- The movements performed while kettlebell training tend to mimic real-life movements, making it an extremely functional form of exercise.
- Utilizing kettlebells helps generate a full-body workout, as you are using multiple muscle groups and joints that will benefit your overall strength and endurance.

Let's explore the anatomy of a kettlebell, ways kettlebell training can help athletes in a variety of sports, and hand care.

Anatomy of a Kettlebell

A kettlebell looks like a cannonball with a handle attached to it. Kettlebells come in all sizes, from a petite 4.5-kilogram (10 lb) version to a very large 92 kilograms (203 lb). The handle and the body of the kettlebell are all one piece, so there is no assembly required.

One of the training advantages of kettlebells is the difference in size between the handle and the body. The kettlebell's offset center of gravity helps promote mobility and stability, especially during overhead lifts. A kettlebell is not like a dumbbell or barbell where the weight is evenly distributed on both ends of the handle or where you put your hands. The asymmetrical design of the kettlebell will force you to work harder to accomplish the exercise you are performing. Kettlebell training will crank up your upper-body strength and overall resilience.

Kettlebell Training for Sports

Kettlebells work the body in similar ways to barbells, sandbags, and other free weights. Many sports, athletic activities, and everyday activities such as housework or gardening benefit from a stronger body overall. It is difficult to select a specific sport or sports that benefit from certain kettlebell exercises, as athletes in literally all sports can benefit from the neuromuscular enhancement kettlebell training provides.

Hand Care

Any work or exercise during which you constantly grip or grab something can cause calluses on your hands. Make sure you manage these calluses often so you can continue to exercise and perform activities of daily life. It goes beyond the intent of this book to tell you how to take care of your hands, but if you search "kettlebell hand care," you should be able to find everything you need to know. Lifting gloves are not recommended because they tend to dull or decrease the ability of the nerve receptors in the hands. These receptors feed back to our brain what our hand is doing and help generate the reactive nerve impulses as they go to the rest of our body based on what our hand is doing.

UNDERSTANDING MUSCLES

The human body contains three types of muscle: cardiac, smooth, and skeletal. In this book, we'll be primarily dealing with the skeletal muscles that were designed to move us through space. Cardiac muscle, also known as myocardium, is a striated muscular tissue exclusive to the heart. It is involuntary regarding its actions and is entirely responsible for the heart's capacity to pump blood throughout the body. One of the actions of cardiac muscle is to move deoxygenated blood to the lungs and, upon the blood's return to the heart, to move oxygenated blood to the body. Smooth muscle is nonstriated, not organized, generally found in our visceral organs, and involuntary.

A tendon is made of connective tissue that connects muscle to bone. Upon a signal from the nerve that attaches to the relative muscle, the muscle will contract and the bone it is connected to will also move through the strong, flexible fibrous connection of the tendon. Additionally, the tendon also protects the muscle from injury when you run, jump, or perform other movements by reducing the impact the muscle would receive.

Muscles and tendons generally are thought of separately, but they are intertwined in an intimate sense. A skeletal muscle, for example the biceps brachii (figure 1.1), includes not only the muscle fibers but also the tendons at either end that are connected to bone. When you lift something from your waist to your mouth, you use your biceps muscle. The muscle fibers of the biceps brachii contract and relax to move the bones they are attached to, via the tendons, in a voluntary way based on the action you are trying to achieve. Tendons do not generally contract and relax as muscle fibers do, but they are much stronger in their tensile strength. Finally, the nerves attached to the muscle fibers tell the muscles to contract or relax. This neuromusculoskeletal relationship throughout the body allows us to get up from a chair, walk across the room, pick up a glass, and bring it to our mouth to have some water.

There are two types of skeletal muscle fibers: fast twitch and slow twitch. Fast-twitch fibers contract very quickly and exhibit a large amount of power but tend to fatigue quickly. For example, when a track and field athlete runs a 100-meter dash, their fast-twitch fibers are primarily being utilized. Slow-twitch fibers can contract for long periods of time and generate a lower level of power when compared to the fast twitch. Examples of this type of fibers are postural muscles that can allow us to stand for hours or to simply hold our head up for a time.

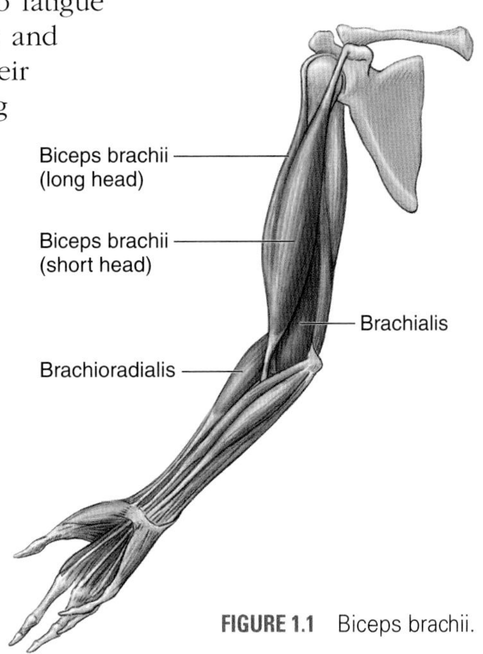

FIGURE 1.1 Biceps brachii.

The exercises in this book will help you develop both types of skeletal muscles in varying degrees. For example, consider the performance of the kettlebell swings in chapter 3. While the fast-twitch fibers of the gluteus maximus and hamstrings will definitely be enhanced, the muscles of your lower leg, especially the gastrocnemius and soleus muscles, will be utilized to maintain your posture while performing multiple repetitions of the kettlebell swing.

ADVANTAGES OF KETTLEBELL TRAINING

Kettlebell training provides key advantages in the development of strength, endurance, mobility, and overall conditioning.

Strength

When using kettlebells in your training, it is difficult to isolate specific muscles, as training with kettlebells will enhance your body in a global sense. For example, when you perform a kettlebell press as in chapter 4, everything from your rib cage down will be strongly activated in an isometric contraction to maintain your posture while you dynamically press the kettlebell overhead. Your pressing muscles, primarily your deltoid and triceps muscles, are working hard, and your abdominal and gluteal muscles are maintaining the isometric posture in a strong contraction. Another example is the renegade row, discussed in chapter 8, which is one of my personal favorite exercises. In this exercise, while in the top of a push-up position, you row a weight with your right hand, then with your left hand, and repeat for the set. While doing this, your hips do not move, and neither do your legs. Moving from four points of contact (both feet and both kettlebells are touching the ground) to three points of contact (both feet and only one kettlebell is touching the ground while lifting one kettlebell in the air) and back to four points of contact is what makes this exercise fantastic. Maintaining this posture not only makes the exercise harder and more beneficial, but it also trains the appropriate muscles and decreases the compensatory mechanisms that a lot of people use to accomplish a task.

Mobility

A lot of people think of strength, not mobility, when they think about kettlebells. However, kettlebells can be used to train both strength and mobility. From a strength perspective, almost every exercise in this book can make you stronger and healthier. But using kettlebells to enhance mobility has transformed my training and has benefited a lot of my patients to be able to move better and decrease their injury potential at the same time. The kettlebell arm bar, in chapter 10, is a great example in that when performing this exercise, your thoracic spine and shoulder mobility will increase over time. Some people try to use a significant amount of weight when performing the arm bar, but in my professional opinion they are missing the boat when performing it this way. Personally, I tend to use either the 12-kilogram or 16-kilogram (35 lb) kettlebell while doing this mobility movement. Another one of my favorite kettlebell mobility movements is the prying goblet squat in chapter 6. This particular mobility movement is paramount for anyone doing any type of squatting motion or single-leg movement exercise. I also recommend a light weight, such as a 12-kilogram (26 lb) to a 20-kilogram (44 lb) kettlebell, as you are not trying to gain strength but instead enhance your hip and pelvic mobility prior to your training session.

Performed on a regular basis, these mobility movements will decrease the negative physical stress that our bodies are under in our society, especially with the advent of electronic devices used nearly 24 hours, 7 days a week.

Excess Postexercise Oxygen Consumption

Excess postexercise oxygen consumption (EPOC) refers to the increase in metabolism (the rate at which calories are burned) after a training session. After a kettlebell snatch, for example, your metabolism will be increased partly due to the increased amount of oxygen that was consumed. EPOC is one of the positive side effects of high-intensity exercise. For those persons wanting to lose weight, EPOC can definitely help. Understanding EPOC is important especially when you are performing ballistic kettlebell exercises and a few grind exercises, too. Even when you are done exercising, you are still burning calories afterward—even while resting.

PROPER TECHNIQUE

Successful kettlebell training means not only performing exercises with proper technique but also being aware of your surroundings while training to ensure both safety and efficiency. Many of the safety guidelines required for safe training with any implement—dumbbells, barbells, machines, and sandbags—also apply to kettlebells. But kettlebells also offer unique challenges and benefits.

Safety Considerations

As with all strength training implements, always keep safety in the back of your mind when lifting. Being safe also means being strong in both mind and body to be able to continue for the next training session.

Before beginning kettlebell training, or any training, obtain medical clearance. See your health care provider and ask them if you are healthy enough to start working with kettlebells. If you have any heart-related health issues, see your cardiologist before starting.

Maintain awareness of your surroundings. Make sure the area where you are training is clear of any equipment, so you have a safe area in which to train. The size of the area needed depends on the exercise: A kettlebell swing requires more room than a squat, for example. Make sure the floor is not slippery.

Train barefoot or wear shoes that have a flat, thin sole and space for the toes to stretch. Our feet have thousands of nerve receptors that communicate northward to our brain. Performing kettlebell training in bare feet enhances these nerve receptors and benefits our balance and coordination. Work boots and thick-soled athletic shoes diminish the ability of these receptors to work for us while we perform kettlebell training. If you do need to wear shoes, choose a shoe with a zero toe-to-heel rise.

Work with the kettlebell, not against it. For example, if while doing a get-up or pressing a heavier kettlebell overhead, the repetition goes sideways, don't try to save it; guide the kettlebell to the ground as best as you can. Make sure you move your body and your feet out of the way as quickly as possible.

Practice all safety measures at all times. From the moment you pick up the kettlebell to the time you are performing the exercise and back to putting the kettlebell away, make sure that you practice safety. You do not want to do a hard and near-perfect set of 10 repetitions of a kettlebell swing and then use poor form picking a kettlebell up to put it away. That is a good way to get injured.

Focus on quality, not quantity. This is a proverb all my kettlebell students know by heart. Quality is always more important than quantity, especially as fatigue starts to set in while performing either a ballistic or a grind exercise. Quantity is also an important aspect in reference to the volume that we are trying to achieve. Doing four high-quality repetitions and stopping the set at that point and not continuing is better than doing a fifth repetition with poor quality. Our neuromuscular system tends to remember the last repetition as the way to perform when we are fatigued. Never train to failure. This is where most injuries occur.

When your heart rate is up, keep moving. As with any exercise that increases your heart rate, keep moving when resting as your heart rate decreases, even at a slow pace. This continues to move blood around and helps your heart circulate it throughout your body.

Maintain the spine in a neutral posture during exercise. As I teach my students to lift properly, one of the key areas I focus on is making sure that their spine is in a neutral position, from the base of the skull to the tailbone. During the movement and while lifting a load, the hips and shoulders, not the spine, will be moving and performing the exercise.

Gradually increase the workout load, employing common sense and listening to your body. While you may feel like a superhero, use common sense and gradually build up the weight, sets, and repetitions that you use during your training. *Consistent* and *persistent* are two very important words to use when strength training. In addition, don't push yourself into positions that you are not accustomed to. Build up your flexibility and mobility. Finally, instruction cannot cover all scenarios. There is no replacement for sound judgment.

Grips: Pronated, Supinated, or Neutral

Imagine you are sitting in a chair. If the palms of both hands are on your knees, your hands are pronated. Supination is the opposite; your hands are palms up. In a neutral position, your palms face each other, thumbs pointing up.

Hip-Hinge Position and Hip Snap

A kettlebell swing starts with the kettlebell on the ground. Before moving the kettlebell, the ideal hip position for the kettlebell swing is a hip hinge, which is when your hips are above your knees and below your shoulders. Your tailbone would be pointing at the 8 o'clock position on an analog clock. This would be the same position your hips would get into if I ask you to jump as high as possible and you did a countermovement before jumping.

Hip snap is the movement from a hip hinge to the vertical plank and what occurs specifically at the hip joint. It ends up in the vertical plank position at the top of the kettlebell swing (figure 1.2). As you move to this position, imagine snapping your hips into position. This requires a maximal contraction of your gluteus maximus, the strongest and biggest muscle in our bodies.

FIGURE 1.2 Kettlebell swing: *(a)* hip hinge; *(b)* hip snap.

Swing Switch

When performing a one-handed swing inside the legs, switching the kettlebell to the other hand can be done one of two ways. The first way is called the swing switch. As you swing the kettlebell up to where your arm is approximately parallel to the ground, bring your other hand up toward the kettlebell, grab the handle while at the same time letting go with the other hand, and then continue swinging the kettlebell. The other way to switch hands is to set the kettlebell down on the ground, let go of the handle, and then grab it with the other hand and start to swing again. The first method, the swing switch, is preferred. The swing switch is used not just for kettlebell swings, but for any movement that requires you to switch hands and sides and do it quickly.

Asymmetrical Loading

In asymmetrical loading, you place an uneven load on your body as you train. Kettlebells are especially useful for asymmetrical loading. Often, asymmetrical loading is not looked upon favorably. However, this method of loading can be beneficial if done properly.

Performing asymmetrical loading means acting as if you are loading your body evenly, even though you are not, as you move your body through space. While performing the exercise with asymmetrical loading, you will contract certain muscles on one side of your body more than the same muscles on the other side of the body due to the uneven load.

Antirotation, antilateral flexion, antiextension, and antiflexion exercises are all forms of asymmetrical loading. Once you learn to properly perform symmetrical loading exercises (e.g., two-handed swing), progress to the asymmetrical loading exercises (e.g., one-handed swing). In the kettlebell swing, you will find that the contralateral gluteus medius, gluteus minimis, and quadratus lumborum will contract harder than the same ipsilateral muscles. If the asymmetrical swing is practiced on both sides of the body, you will find that your two-handed swing becomes stronger.

Breathing

Proper breathing while performing any exercise, regardless of implement, provides additional benefits, including better focus and reduced risk of injury. With kettlebell training, two key aspects of proper breathing are breathing behind the shield and the biomechanical breathing match.

Breathing Behind the Shield

When picking up a weight, you must brace your abdominal muscles to protect your spine and transfer the forces to the proper areas to be successful in your lifting. Breathing behind the shield (i.e., your abdominals) helps to do this, not only in the grind exercises but also in the ballistic ones.

A grind exercise, such as the get-up (chapter 5) or the squat (chapter 6), usually refers to a long, physically draining movement of moderate difficulty and moderate to high intensity. Movement speed is low, mass is high. A ballistic exercise, such as the kettlebell swing (chapter 3), alternates between tension and relaxation during the swing. Performing the swing in this manner teaches a person to generate large forces and then immediately relax and then repeat—tension and relaxation; back and forth. Movement speed is high, mass is moderate.

To picture breathing behind the shield, imagine lying on your back and having someone stand on your abdominal muscles. To protect your organs from being crushed, you would strongly contract your abdominal muscles, also known as abdominal bracing. Additionally, to be able to continue talking, your breaths would be shallow but enough for you to continue. If you have any cardiovascular health issues, please consult your physician before attempting this type of training.

Biomechanical Breathing Match

Biomechanical breathing match is the syncing up of your inhalation and exhalation with the exercise movement. Most strength exercises have a negative, eccentric component coupled with a positive, concentric component.

An eccentric contraction occurs when the primary muscle is lengthening under load, such as when you are descending in the squat (chapter 6). A concentric contraction is when the primary muscle is shortening under load, as in the ascension of the squat.

For most exercises, you will breathe in during the eccentric motion and breathe out on the concentric aspect. This is a biomechanical breathing match. I recommend that you incorporate a few other factors. First, breathing in through your nose strongly activates your diaphragm, and you get a better breath. Second, as you are inhaling, imagine inhaling into your groin. Your lung tissue does not go past your diaphragm, but as you breathe diaphragmatically, you will feel it all the way down to your pelvic floor. Third, during the concentric portion of the lift, do not exhale completely. If you do, you will lose the intraabdominal pressure that protects your spine and was created during abdominal bracing. Instead, hiss powerfully at the top and keep tension in your abdomen while maintaining the bracing at the top. See breathing behind the shield for more information about this.

Taking a Photo

If I came into the gym when you were at the top of the get-up, press, or kettlebell snatch, I should not be able to tell which exercise you were performing. I love this about kettlebell training; three different movements end in the same position, but within a second, they are three different exercises again. The moment you move the kettlebell, I could tell which exercise you were performing.

If I took a still photograph of any part of the get-up, I should not be able to tell if you are ascending or descending with the kettlebell. When performing this movement, maintain your form and technique on both the ascent and the descent.

Water Bowl

When performing a kettlebell movement—for example the double kettlebell front squat—the movement is primarily a hip, knee, and ankle movement. The spine, which starts at the base of the skull and goes down to the bottom of the tailbone, should not move when you are squatting with load. If you drop your pen on the floor and bend over to pick it up, how you bend your body does not matter. If you are squatting double 24-kilogram (53 lb each) kettlebells, what you do with your spine is crucial.

As you go from a vertical plank to a squat below parallel (the hips are below the knees) and back to the vertical plank position, your rib cage should remain stacked on top of your pelvis. Imagine your pelvis is a bowl filled with water (figure 1.3). If you anteriorly tilt it forward or posteriorly tilt it backward, water will spill out. Your rib cage should stay horizontal on top of this bowl. If you anteriorly tilt the bowl and spill water and the front part of the rib cage tilts up, your lower spine, namely your lumbar region, is extended and compressed. This could lead to future health problems for your lower spine. In other words, don't spill water from the bowl.

Antishrugs

To shrug or not to shrug, that is the question!

To steal a line from William Shakespeare, we will discuss the movement known as the antishrug. Throughout this book, you will see the word *antishrug* listed in various exercises to help promote not only proper shoulder health but also increased stabilization for both the exercise that you are performing and your spine. To perform the antishrug movement, I teach my students to first shrug their shoulders and then perform the opposite—antishrug their shoulders. Try and touch your fingers to the ground while standing tall. Keeping your neck on top of your shoulders while performing the antishrug movement is paramount to proper health and mechanics of your upper body and neck regions.

By performing the antishrug movement, you are contracting your latissimus dorsi, among other muscles. This creates a co-contraction of your abdominals, which further increases the stabilization of your spine through the thoracolumbar fascia, which attaches to the lumbopelvic region. When you antishrug prior to your exercises, I can almost guarantee that you won't round your lower back. This muscle also depresses your scapula, which further packs your shoulders. It also puts the glenohumeral (shoulder) joint in a safe and packed position, especially when lifting heavy weight. These important qualities of your latissimus dorsi muscles will help to decrease injuries and increase your

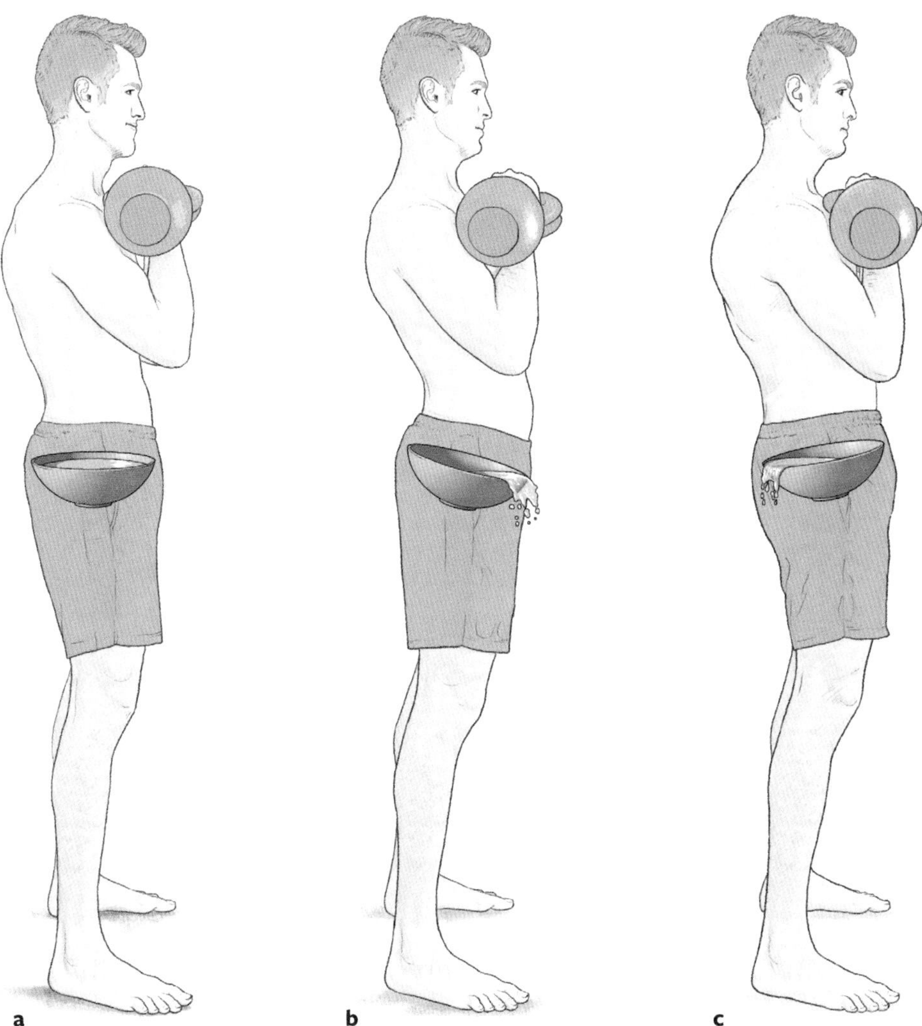

FIGURE 1.3 Double kettlebell front squat with water bowl: *(a)* proper position; *(b)* anterior pelvic tilt; *(c)* posterior pelvic tilt.

performance, both in the gym and outside of it. To answer the question that started this description, don't shrug but antishrug!

Exercise Ratings

At the beginning of each exercise in this book, you will see one to three kettlebells present. These kettlebells represent the level of difficulty of each exercise: One kettlebell indicates a beginner exercise, two kettlebells indicate an intermediate exercise, and three kettlebells indicate an advanced exercise. Please use this rating when choosing which exercises to perform as you progress with your strength training.

CONCLUSION

Now that you have read this chapter, go back and read it again. The following chapters describe the proper technique for many kettlebell movements. These exercises will not only increase your strength but will also lead to better health overall. There are many variations of the basic movements of kettlebell training, however, I recommend spending a good deal of time learning to do the basic movements first and then start incorporating variations, if needed. Someone once said the main difference between a novice athlete and an elite one is that the elite athlete can do the basics or fundamentals much better. Chapter 2, Deadlift, will get you started in learning the fundamentals of lifting with kettlebells as well as lifting in general. Many of the principles covered throughout this book, including this chapter, are applicable when using other implements such as barbells, dumbbells, or sandbags. Have a strong day!

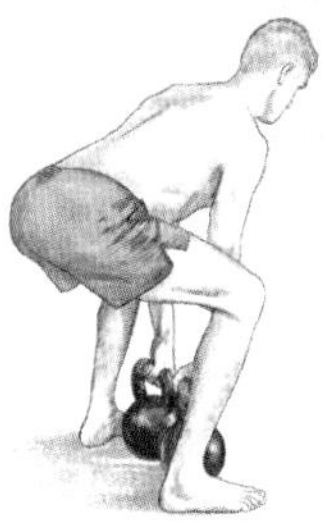

2

DEADLIFT

Whole-body tension. Strength. Lats. This chapter, the first in which we feature proper kettlebell technique, will be about performing the kettlebell deadlift. Before you learn the swing or the squat, you must become proficient with the deadlift. The basic action of the deadlift is the same regardless of the implement being used—kettlebell, barbell, sandbag, trap bar, or dumbbells. They are all used to make someone stronger, especially for motions involving picking something off the ground. The deadlift allows you to generate total-body tension to be able to lift the object off the ground, not preceded by an eccentric contraction. It is also a very simple movement that has been used for centuries. This movement has great applications across all sports that require the athlete to go from a dead stop to a strong movement.

When performing deadlifts with a kettlebell, there will be a limitation on how much weight you can use because of the nature of the kettlebell itself. They look like cannonballs with handles—a piece of iron where the weight is already chosen; if you wish to either increase or decrease the weight, choose a different size kettlebell.

One of the benefits of learning kettlebell deadlifts is learning how to perform a vertical, or standing, plank while loaded. This is essential for you to understand and perform as we progress through the following chapters. Another benefit is learning how to turn on the proper tension that is needed to lift the kettlebell prior to it leaving the ground.

One of the analogies that I like to use to teach the kettlebell deadlift is that of airline pilots. Pilots go through a checklist before every flight. When they were learning how to fly, it took them longer to go through the checklist. Once they are seasoned pilots, they still go through the checklist but do so much faster. The same principle applies to the kettlebell deadlift. At first it will take a while to make sure everything is set right before you lift the kettlebell. Eventually it will take you only one to three seconds before each repetition to go through your mental checklist.

Once you learn the proper technique and tension required to perform the deadlift, you will see improvements in your posture, better strength in picking objects up from the ground, and an overall positive change in your ability to use your body in athletic and recreational hobbies, such as basketball (figure 2.1).

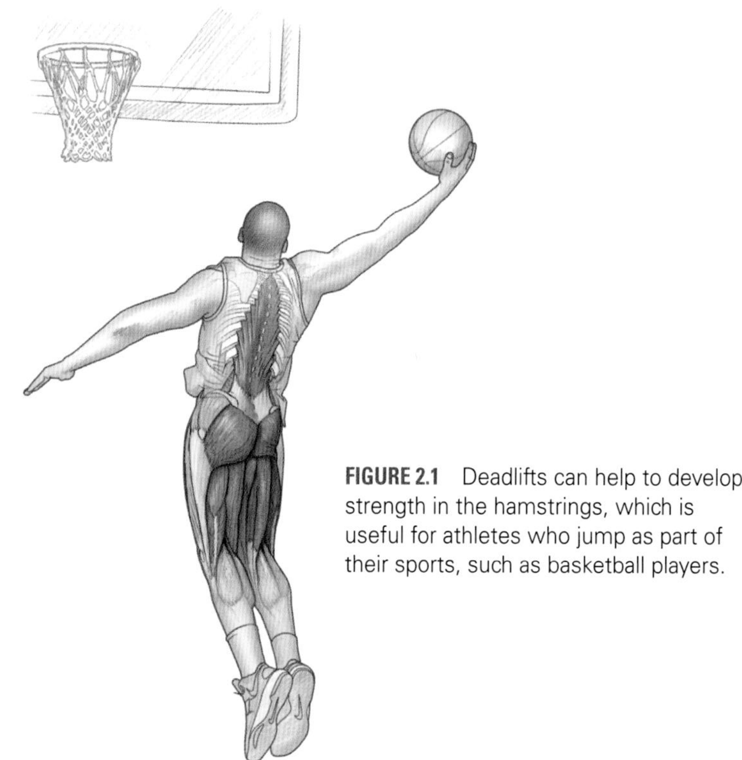

FIGURE 2.1 Deadlifts can help to develop strength in the hamstrings, which is useful for athletes who jump as part of their sports, such as basketball players.

EXERCISE FOCUS FOR KETTLEBELL DEADLIFTS

The exercises in this chapter can be organized into three categories:

1. Kettlebell deadlift, inside legs
2. Kettlebell deadlift, outside legs
3. Single-leg deadlift

Each category of deadlift provides different ways to help you learn proper technique and develop strength. Here, we take a closer look at the focus of the exercises in each category.

Kettlebell Deadlift, Inside Legs

Deadlifting with a kettlebell inside your legs is generally the easiest way to learn the proper form of deadlifting. Some of the advantages that you will gain from performing it this way are that you will be able to create and hold a vertical plank position at the top of the lift; you will learn proper tension-developing skills, which will be used later with other strength movements; and you will learn the proper setup for the kettlebell swing and other related movements. In addition, the single kettlebell one-hand deadlift also further trains the side of your body that does not have a hand on the kettlebell to work hard and maintain a straight spine laterally throughout the movement. With the double kettlebell deadlift, you will also learn to handle a kettlebell in each hand, which will be useful when learning a double kettlebell swing and clean. Most people find it easier to set up for the kettlebell deadlift inside the legs with the feet between hip- and shoulder-width apart.

Kettlebell Deadlift, Outside Legs

A deadlift with the kettlebell outside your legs is performed like a conventional deadlift with a barbell. However, unlike the barbell position, the hips and knees are flexed approximately the same for the kettlebell deadlift outside the legs. Similar to deadlifting with the kettlebell inside the legs, this exercise teaches you to hold a vertical plank position at the top of the lift. Furthermore, while performing the double kettlebell deadlift and using heavier weight, you will learn to develop proper tension, which will be used with other strength movements, and you will be handling a kettlebell in each hand, balancing the overall load. The single kettlebell deadlift with one hand, outside the legs, also known as the suitcase deadlift, teaches you to lift an asymmetrical load while keeping your spine, hips, and pelvis centered. The side of your body not holding the kettlebell will work hard to maintain a straight spine since the kettlebell is farther laterally from the center of your body.

Single-Leg Deadlift

The single-leg deadlift is an interesting exercise movement in that even though it uses a unilateral approach, it improves the entire body. Some of the benefits from performing the single-leg deadlift include being able to dynamically balance on one leg, thereby improving overall balance and stability. You will also find that it is usually easier to balance on one leg than the other one. Keep practicing to make the not-so-strong leg better. It also teaches proper tension, developing skills in an asymmetrical way that will be used with other strength movements. Adding a kettlebell to this exercise enhances the balance and strength of the stance leg by adding weight. A further progression is to hold a kettlebell in each hand, increasing the load and further enhancing the balance and strength of the stance leg.

SINGLE KETTLEBELL DEADLIFT, INSIDE LEGS

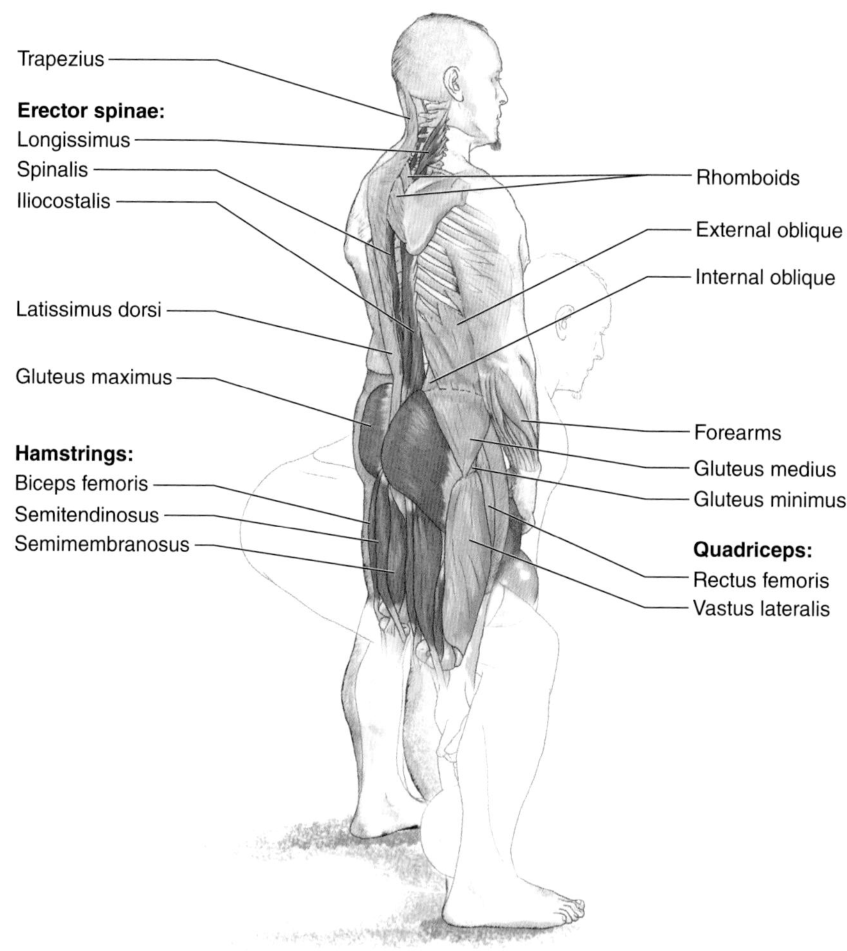

Execution

1. Take a shoulder-width stance over the kettlebell. Angle the toes slightly outward. Position the handle just in front of the legs.
2. Keeping the hips above the knees and below the shoulders, grip the kettlebell handle with both hands, placing your body into a hip-hinge position. You should feel tension in your hamstrings.
3. Grip the ground with your toes. Tighten the latissimi dorsi, the abdominal muscles, and your grip. Stand up with the kettlebell until you are fully erect, squeezing your gluteal muscles hard.
4. Repeat the same path on the way down. Reset and repeat.

Muscles Involved

Primary: Erector spinae (iliocostalis, longissimus, spinalis), gluteus maximus, hamstrings (semitendinosus, semimembranosus, biceps femoris)

Secondary: Trapezius, rhomboids, latissimus dorsi, quadriceps (rectus femoris, vastus lateralis, vastus medialis, vastus intermedius), gluteus medius, gluteus minimus, rectus abdominis, external oblique, internal oblique, transversus abdominis, forearms (wrist flexors, finger flexors)

Anatomic Focus

Hand spacing: Place both hands next to each other on the handle of the kettlebell. Imagine trying to break the handle, thereby strongly activating the latissimus dorsi muscle.

Grip: Use a double overhand grip (pronate both hands).

Stance: Place the legs approximately shoulder-width apart with the kettlebell between the legs. Angle the toes slightly outward. Position the knees at 60 to 70 degrees of flexion.

Trajectory: Maintain a kettlebell path that is straight up and down and close to the body.

Range of motion: Keeping your arms extended and your elbows stiff, stand up with the kettlebell. Keep your spine straight and stiff throughout the movement. The erector spinae, abdominal muscles, and latissimus dorsi muscles will help to stabilize and straighten the spine while the gluteus maximus and hamstrings generate hip extension. Antishrug your shoulders (depress the shoulders away from your ears versus the opposite of shrugging/elevating the shoulders toward your ears), further activating your latissimus dorsi muscles. Tighten your gluteal muscles to stand straight up. Do not overextend your spine.

DOUBLE KETTLEBELL DEADLIFT, INSIDE LEGS

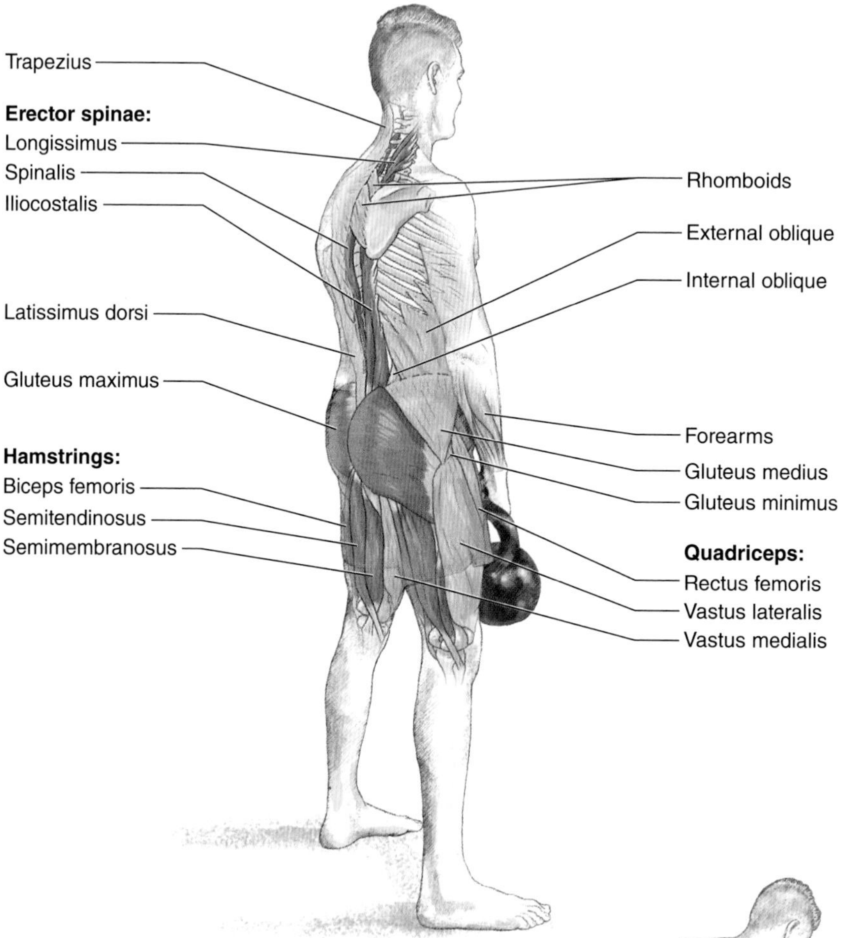

Trapezius

Erector spinae:
Longissimus
Spinalis
Iliocostalis

Latissimus dorsi

Gluteus maximus

Hamstrings:
Biceps femoris
Semitendinosus
Semimembranosus

Rhomboids

External oblique

Internal oblique

Forearms

Gluteus medius

Gluteus minimus

Quadriceps:
Rectus femoris
Vastus lateralis
Vastus medialis

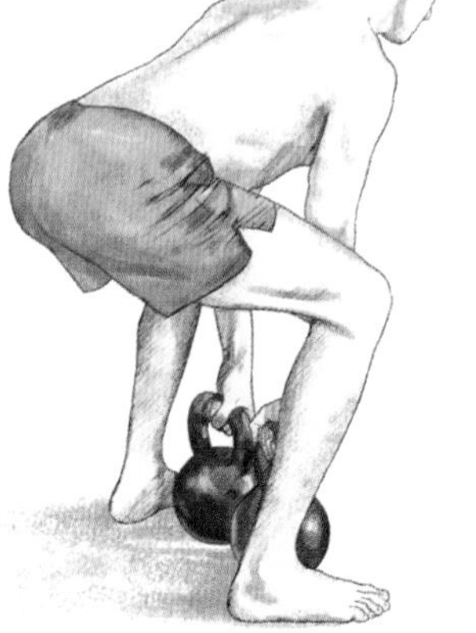

Starting position.

Execution

1. Take a slightly wider than shoulder-width stance over the kettlebells. Angle the toes slightly outward. The kettlebell handles should be just in front of the legs.
2. Keeping the hips above the knees and below the shoulders, grip a kettlebell handle in each hand, placing your body into a hip-hinge position. You should feel tension in your hamstrings.
3. Grip the ground with your toes. Tighten the latissimi dorsi, the abdominal muscles, and the grip. Stand up with the kettlebells until you are fully erect, squeezing your gluteal muscles hard.
4. Repeat the same path on the way down. Reset and repeat.

Muscles Involved

Primary: Erector spinae (iliocostalis, longissimus, spinalis), gluteus maximus, hamstrings (semitendinosus, semimembranosus, biceps femoris)

Secondary: Trapezius, rhomboids, latissimus dorsi, quadriceps (rectus femoris, vastus lateralis, vastus medialis, vastus intermedius), gluteus medius, gluteus minimus, rectus abdominis, external oblique, internal oblique, transversus abdominis, forearms (wrist flexors, finger flexors)

Anatomic Focus

Hand spacing: Put each hand on its own kettlebell handle.

Grip: Use a double overhand grip (pronate both hands).

Stance: Position the legs slightly more than shoulder-width apart with the kettlebells between the legs. Angle the toes slightly outward. Flex the knees.

Trajectory: Use a kettlebell path that is straight up and down and close to the body.

Range of motion: Keeping your arms extended and your elbows stiff, stand up with the kettlebells. Keep your spine straight and stiff throughout the movement. The erector spinae, abdominal muscles, and latissimus dorsi muscles will help to stabilize and straighten the spine while the gluteus maximus and hamstrings generate hip extension. Antishrug your shoulders, further activating your latissimus dorsi muscles. Squeeze your gluteal muscles to stand straight up. Do not overextend your spine.

SINGLE KETTLEBELL DEADLIFT WITH ONE HAND, INSIDE LEGS

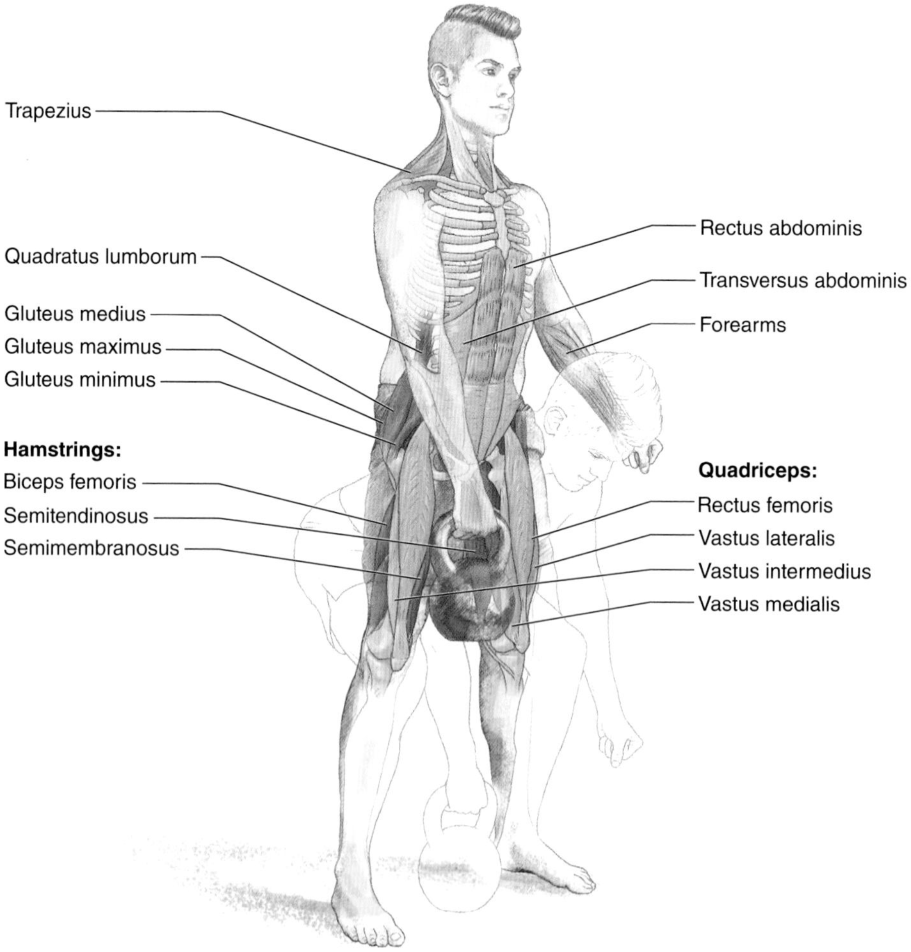

Trapezius

Quadratus lumborum

Gluteus medius

Gluteus maximus

Gluteus minimus

Hamstrings:

Biceps femoris

Semitendinosus

Semimembranosus

Rectus abdominis

Transversus abdominis

Forearms

Quadriceps:

Rectus femoris

Vastus lateralis

Vastus intermedius

Vastus medialis

Execution

1. Take a shoulder-width stance over the kettlebell. Angle the toes slightly outward. The handle is just in front of the legs.
2. Keeping the hips above the knees and below the shoulders, grip the kettlebell handle with one hand, placing your body into a hip-hinge position. Keep the other hand to the side of the handle but not on the handle. You should feel tension in your hamstrings.
3. Grip the ground with your toes. Tighten the latissimi dorsi, the abdominal muscles, and your grip. Stand up with the kettlebell until you are fully erect, squeezing your gluteal muscles hard. Resist the urge to let your body bend sideways while standing up.
4. Repeat the same path on the way down. Reset and repeat. After completing all repetitions with one hand, switch hands.

Muscles Involved

Primary: Erector spinae (iliocostalis, longissimus, spinalis), gluteus maximus, gluteus medius, gluteus minimus, hamstrings (semitendinosus, semimembranosus, biceps femoris), quadratus lumborum

Secondary: Trapezius, rhomboids, latissimus dorsi, quadriceps (rectus femoris, vastus lateralis, vastus medialis, vastus intermedius), rectus abdominis, external oblique, internal oblique, transversus abdominis, forearms (wrist flexors, finger flexors)

Anatomic Focus

Hand spacing: Place one hand on the handle of the kettlebell. Place the other hand next to the side of the handle, but not on it, to help with proper body and torso position at the beginning and throughout the movement.

Grip: Use a single overhand grip (pronate the hand on the kettlebell).

Stance: Position the legs approximately shoulder-width apart with the kettlebell between the legs. Angle the toes slightly outward and slightly flex the knees.

Trajectory: Use a kettlebell path that is straight up and down and close to the body.

Range of motion: Keeping your arms extended and your elbows stiff, stand up with the kettlebell. Keep your spine straight and stiff throughout the movement. The erector spinae, abdominal muscles, and latissimus dorsi muscles will help to stabilize and straighten the spine while the gluteus maximus and hamstrings generate hip extension. Antishrug your shoulders, further activating your latissimus dorsi muscles. Tighten your gluteal muscles to stand straight up. Do not overextend your spine.

DOUBLE KETTLEBELL DEADLIFT, OUTSIDE LEGS

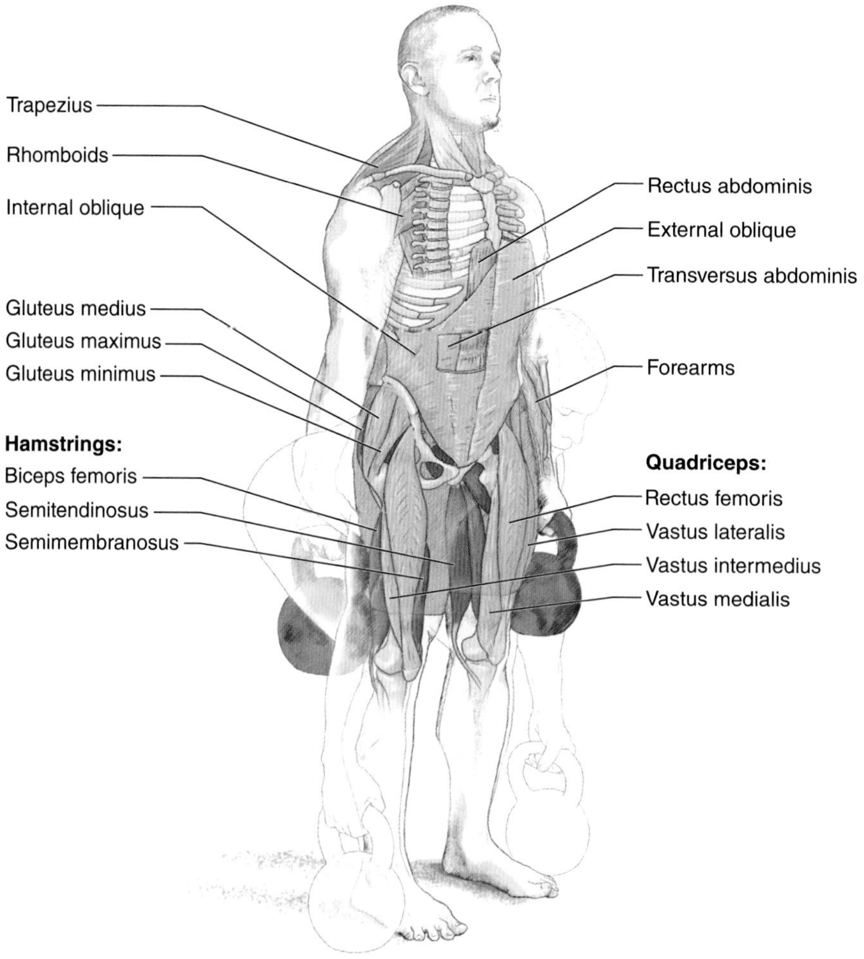

Trapezius
Rhomboids
Internal oblique
Gluteus medius
Gluteus maximus
Gluteus minimus

Hamstrings:
Biceps femoris
Semitendinosus
Semimembranosus

Rectus abdominis
External oblique
Transversus abdominis
Forearms

Quadriceps:
Rectus femoris
Vastus lateralis
Vastus intermedius
Vastus medialis

Execution

1. Take a hip-width stance while placing the kettlebells to the outsides of the ankles. Angle the toes slightly outward.
2. Keeping the hips above the knees and below the shoulders, grip the kettlebell handles with each hand, placing your body into a hip-hinge position. You should feel tension in your hamstrings.
3. Grip the ground with your toes. Tighten the latissimi dorsi, the abdominal muscles, and your grip. Stand up with the kettlebells until you are fully erect, squeezing your gluteal muscles hard.
4. Repeat the same path on the way down. Reset and repeat.

Muscles Involved

Primary: Erector spinae (iliocostalis, longissimus, spinalis), gluteus maximus, hamstrings (semitendinosus, semimembranosus, biceps femoris)

Secondary: Trapezius, rhomboids, latissimus dorsi, quadriceps (rectus femoris, vastus lateralis, vastus medialis, vastus intermedius), gluteus medius, gluteus minimus, rectus abdominis, external oblique, internal oblique, transversus abdominis, forearms (wrist flexors, finger flexors)

Anatomic Focus

Hand spacing: Place each hand on its own kettlebell.

Grip: Place the hands in a neutral grip.

Stance: Position your legs in a hip-width stance with the kettlebells outside the ankles. Angle the toes slightly outward. Slightly flex your knees.

Trajectory: Use a kettlebell path that is straight up and down and close to the body.

Range of motion: Keeping your arms extended and your elbows stiff, stand up with the kettlebells. Keep your spine straight and stiff throughout the movement. The erector spinae, abdominal muscles, and latissimus dorsi muscles will help to stabilize and straighten the spine while the gluteus maximus and hamstrings generate hip extension. Antishrug your shoulders, further activating your latissimus dorsi muscles. Tighten your gluteal muscles to stand straight up. Do not overextend your spine.

SINGLE KETTLEBELL DEADLIFT WITH ONE HAND, OUTSIDE LEGS (SUITCASE DEADLIFT)

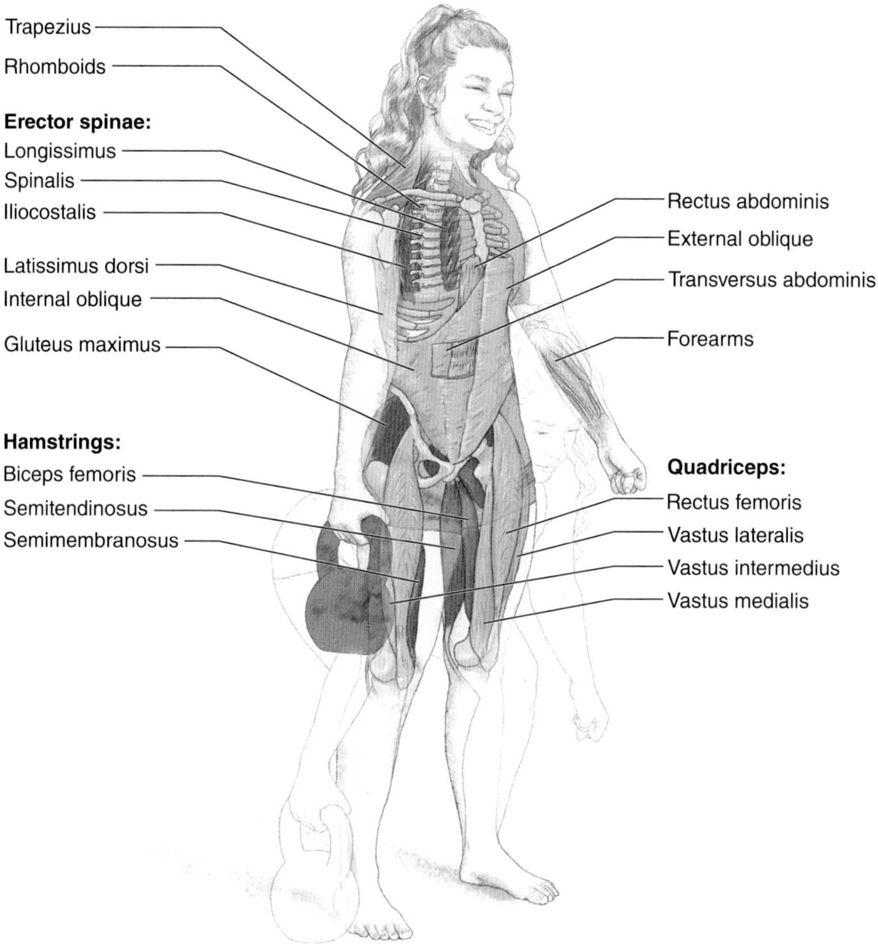

Trapezius
Rhomboids

Erector spinae:
Longissimus
Spinalis
Iliocostalis

Latissimus dorsi
Internal oblique
Gluteus maximus

Hamstrings:
Biceps femoris
Semitendinosus
Semimembranosus

Rectus abdominis
External oblique
Transversus abdominis
Forearms

Quadriceps:
Rectus femoris
Vastus lateralis
Vastus intermedius
Vastus medialis

Execution

1. Take a hip-width stance while placing a kettlebell to the outside of one of your ankles. Angle the toes slightly outward.
2. Keeping the hips above the knees and below the shoulders, grip the kettlebell handle with one hand, placing your body into a hip-hinge position. Keep the other hand to the side of your body in a similar position as if it were lifting a kettlebell. You should feel tension in your hamstrings.
3. Grip the ground with your toes. Tighten your latissimi dorsi, your abdominal muscles, and your grip. Stand up with the kettlebell until you are fully erect, squeezing your gluteal muscles hard. Resist the urge to let your body bend sideways while standing up.
4. Repeat the same path on the way down. Reset and repeat. After completing all repetitions with one hand, switch to the other hand.

Muscles Involved

Primary: Erector spinae (iliocostalis, longissimus, spinalis), gluteus maximus, gluteus medius, gluteus minimus, hamstrings (semitendinosus, semimembranosus, biceps femoris), quadratus femoris

Secondary: Trapezius, rhomboids, latissimus dorsi, quadriceps (rectus femoris, vastus lateralis, vastus medialis, vastus intermedius), rectus abdominis, external oblique, internal oblique, transversus abdominis, forearms (wrist flexors, finger flexors)

Anatomic Focus

Hand spacing: Place one hand on the handle of the kettlebell and the other hand next to the side of your body in a position as though it were lifting a kettlebell. This will help with proper body and torso position at the beginning and throughout the movement.

Grip: Use a single overhand grip (pronate the hand on the kettlebell).

Stance: Place the legs approximately shoulder-width apart with the kettlebell between the legs. Point the toes slightly outward and flex the knees slightly.

Trajectory: Move the kettlebell in a path that is straight up and down and close to the body.

Range of motion: Keeping your arms extended and your elbows stiff, stand up with the kettlebell. Keep your spine straight and stiff throughout the movement. The erector spinae, abdominal muscles, and latissimus dorsi muscles will help to stabilize and straighten the spine while the gluteus maximus and hamstrings generate hip extension. Antishrug your shoulders, further activating your latissimus dorsi muscles. Squeeze your gluteal muscles to stand straight up. Do not overextend your spine.

SINGLE-LEG DEADLIFT WITH BODY WEIGHT

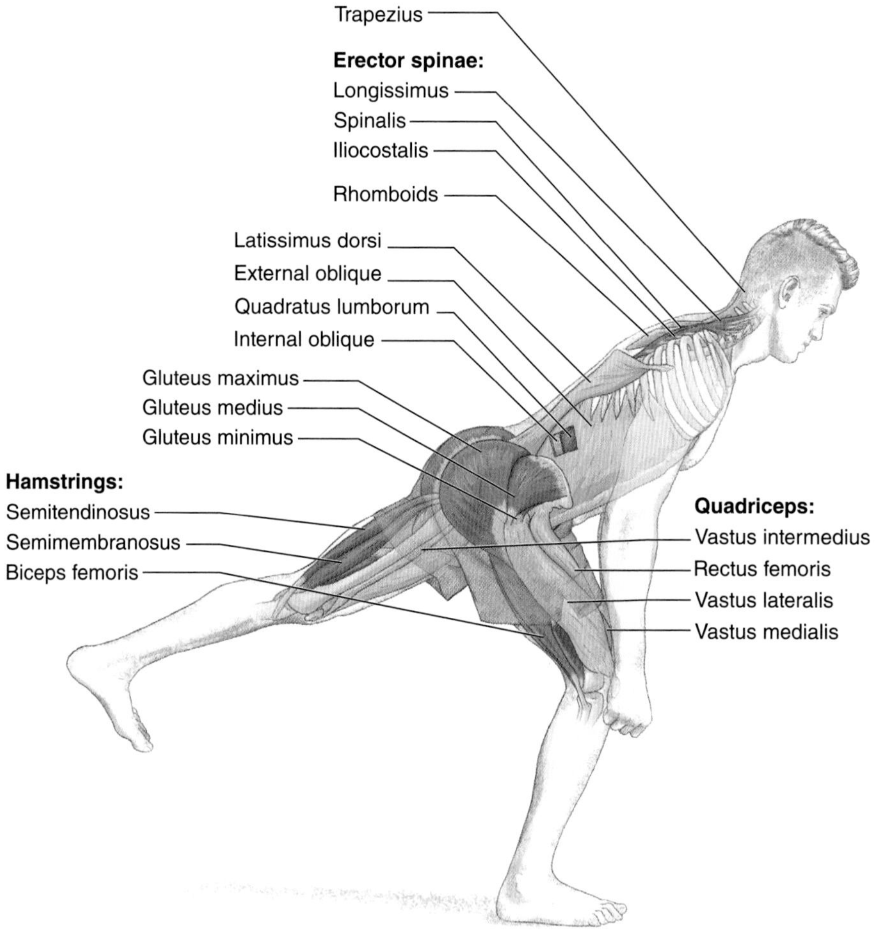

Trapezius

Erector spinae:
Longissimus
Spinalis
Iliocostalis

Rhomboids

Latissimus dorsi
External oblique
Quadratus lumborum
Internal oblique

Gluteus maximus
Gluteus medius
Gluteus minimus

Hamstrings:
Semitendinosus
Semimembranosus
Biceps femoris

Quadriceps:
Vastus intermedius
Rectus femoris
Vastus lateralis
Vastus medialis

Execution

1. Although this exercise doesn't use a kettlebell itself, it is the first step in the progression to performing a single-leg deadlift with a kettlebell. Before adding a kettlebell, make sure you're able to perform this exercise well. Standing on one foot, brace your abdominal muscles and hinge at the hip until you reach full tension in the hamstrings of your standing leg. Make a straight line with your free leg from your ear to your ankle.
2. Grip the ground with your toes. Tighten the latissimus dorsi and abdominal muscles and squeeze the gluteal muscles on your free leg. Dorsiflex the foot of your free leg and point it toward the ground while pushing through the heel.
3. Keep your shoulder and hip level and symmetrical on the eccentric and concentric portion of the lift. Without changing your spinal alignment, try touching the ground with your hands.
4. Drive your stance foot into the ground and keep the bend in your knee between 5 and 15 degrees. Hinge back up to complete the repetition. After completing all repetitions with one leg, switch to the other leg.

Muscles Involved

Primary: Erector spinae (iliocostalis, longissimus, spinalis), gluteus maximus, gluteus medius, gluteus minimus, hamstrings (semitendinosus, semimembranosus, biceps femoris), quadratus lumborum

Secondary: Trapezius, rhomboids, latissimus dorsi, quadriceps (rectus femoris, vastus lateralis, vastus medialis, vastus intermedius), rectus abdominis, external oblique, internal oblique, transversus abdominis

Anatomic Focus

Stance: Use a single leg.

Trajectory: Make a straight line with your free leg from your ear to your ankle. Dorsiflex your foot and point it toward the ground while moving it.

Range of motion: Keep your spine straight and stiff throughout the movement. This is a single-leg hip-hinge movement; therefore the movement should be directed from your hip region. The abdominal and latissimus dorsi muscles will help to stabilize and straighten the spine while the gluteus maximus and hamstrings of the standing leg generate hip extension. Tighten the gluteal muscles on the same side as your standing leg to stand straight up.

(continued)

SINGLE-LEG DEADLIFT WITH BODY WEIGHT *(continued)*

VARIATIONS

Kickstand Deadlift

This exercise is a regression of the single-leg deadlift with body weight. Stand with your feet hip-width apart. Move one foot back until the toes are even with the other heel. Perform this exercise like the single-leg deadlift with body weight, but keep the back foot on its toes. When this exercise gets easier, occasionally lift your back toes off the ground and tap the ground to further load the working muscles of the stance leg. When you are ready, hold a kettlebell in the hand on the side of the back leg.

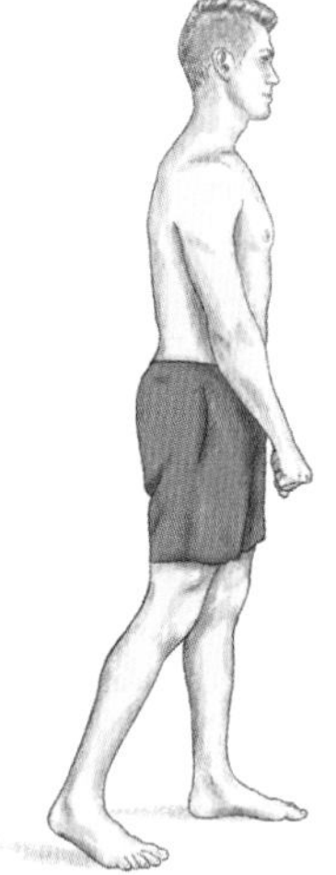

Kickstand deadlift regression, body weight only.

Kickstand deadlift regression with single kettlebell.

Single-Leg Deadlift With Eyes Closed

For this advanced version of the single-leg deadlift with body weight, close your eyes. Move slowly through the range of motion. This version further enhances the training with the stance foot and up the anatomic chain.

SINGLE-LEG DEADLIFT WITH SINGLE KETTLEBELL

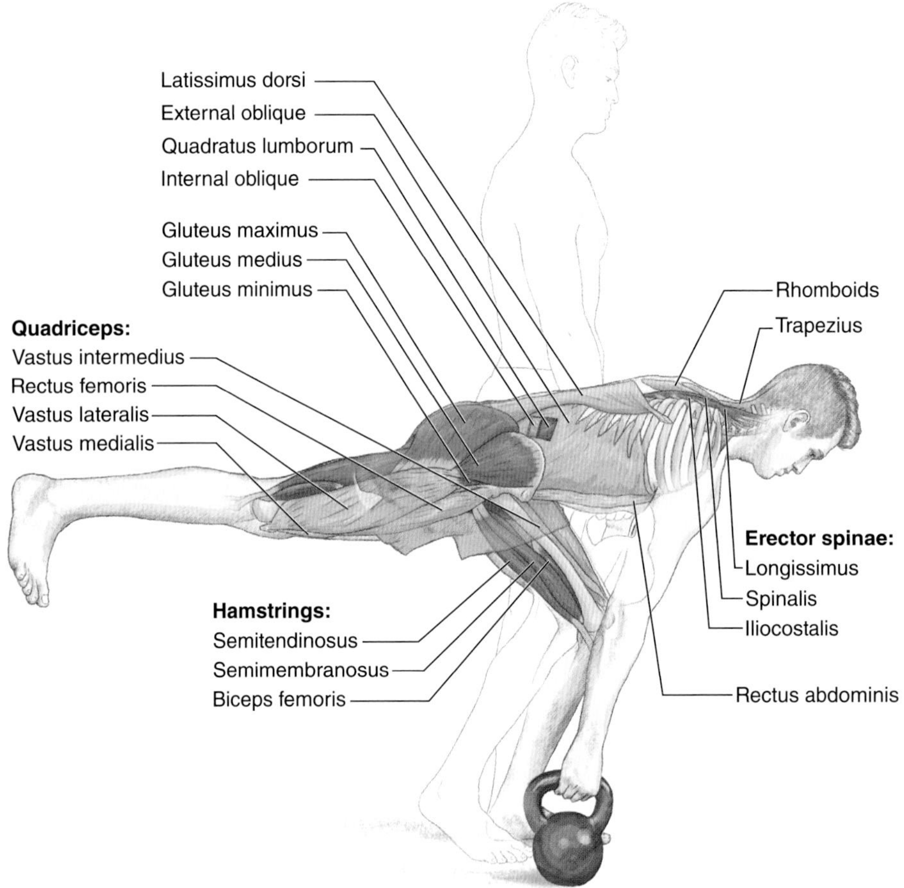

Execution

1. Stand with the kettlebell between your feet. Perform a single-hand deadlift to lift the kettlebell. Your free leg will be the one on the same side as the kettlebell.
2. Place your free leg back and on its toes. Performing the same motion as for the single-leg deadlift with body weight, move your stance leg hip into a hinge position and your free leg back. Make a straight line with your free leg from your ear to your ankle.
3. Grip the ground with your toes. Tighten the latissimus dorsi and abdominal muscles and squeeze the gluteal muscles on the side of your free leg. Dorsiflex your foot and point it toward the ground while pushing through the heel of your lifted leg.

4. Keep your shoulder and hip level and symmetrical on the eccentric and concentric portion of the lift.
5. Drive your stance foot into the ground and keep the bend in your knee between 5 and 15 degrees.
6. Either tap the ground with the kettlebell and stand back up or rest the kettlebell on the ground for a second and then stand up. Complete all repetitions on one leg, then switch to the other leg.

Muscles Involved

Primary: Erector spinae (iliocostalis, longissimus, spinalis), gluteus maximus, gluteus medius, gluteus minimus, hamstrings (semitendinosus, semimembranosus, biceps femoris), quadratus lumborum

Secondary: Trapezius, rhomboids, latissimus dorsi, quadriceps (rectus femoris, vastus lateralis, vastus medialis, vastus intermedius), rectus abdominis, external oblique, internal oblique, transversus abdominis

Anatomic Focus

Hand spacing and grip: Pronate the hand opposite the stance leg and grip the kettlebell with that hand.

Stance: Stand on one leg.

Trajectory: Make a straight line with your free leg from your ear to your ankle. Dorsiflex your foot and point it toward the ground while moving it.

Range of motion: Keep your spine straight and stiff throughout the movement. This is a single-leg hip-hinge movement; therefore the movement should be directed from your hip region. The abdominal and latissimus dorsi muscles will help to stabilize and straighten the spine while the gluteus maximus and hamstrings on the same side as your stance leg generate hip extension. Tighten the gluteal muscles on the same side as your stance leg to stand straight up.

VARIATION

Floor Single-Leg Deadlift With Single Kettlebell

This exercise started from the top, after deadlifting the kettlebell with both legs. You can start from the bottom as well. Move your stance and free leg into position and reach for the kettlebell, keeping your spine in proper alignment. Deadlift the kettlebell up and then set the kettlebell back on the ground, either tapping it or resting it for a short time.

SINGLE-LEG DEADLIFT WITH DOUBLE KETTLEBELLS

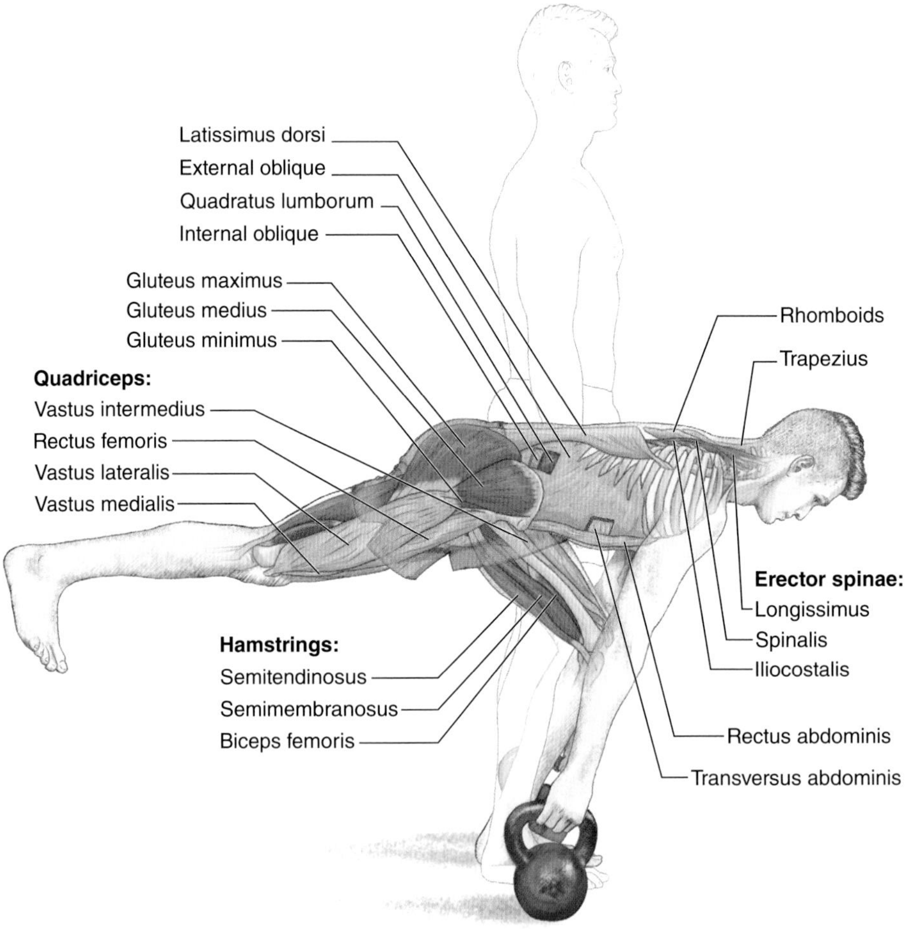

Latissimus dorsi
External oblique
Quadratus lumborum
Internal oblique
Gluteus maximus
Gluteus medius
Gluteus minimus
Quadriceps:
Vastus intermedius
Rectus femoris
Vastus lateralis
Vastus medialis
Rhomboids
Trapezius
Erector spinae:
Longissimus
Spinalis
Iliocostalis
Hamstrings:
Semitendinosus
Semimembranosus
Biceps femoris
Rectus abdominis
Transversus abdominis

Execution

1. Start with the kettlebells between your feet. Perform a double kettlebell deadlift to grip them.
2. Place your free leg back and on its toes. Performing the same movement as the single-leg deadlift with body weight, move your stance leg hip into a hinge position and your free leg back. Make a straight line with your free leg from your ear to your ankle.
3. Grip the ground with your toes. Tighten the latissimus dorsi and abdominal muscles and squeeze the gluteal muscles on your free leg. Dorsiflex your foot and point it toward the ground while pushing through the heel of your lifted leg.

4. Keep your shoulder and hip level and symmetrical on the eccentric and concentric portion of the lift.
5. Drive your stance foot into the ground and keep the bend in your knee between 5 and 15 degrees.
6. Either tap the ground with the kettlebells and stand back up or rest the kettlebells on the ground for a second and then stand up. Complete all repetitions on one leg, then switch to the other leg.

Muscles Involved

Primary: Erector spinae (iliocostalis, longissimus, spinalis), gluteus maximus, gluteus medius, gluteus minimus, hamstrings (semitendinosus, semimembranosus, biceps femoris), quadratus lumborum

Secondary: Trapezius, rhomboids, latissimus dorsi, quadriceps (rectus femoris, vastus lateralis, vastus medialis, vastus intermedius), rectus abdominis, external oblique, internal oblique, transversus abdominis

Anatomic Focus

Hand spacing and grip: Pronate each hand and grip a kettlebell in each hand.

Stance: Stand with a single leg.

Trajectory: Make a straight line with your free leg from your ear to your ankle. Dorsiflex your foot and point it toward the ground while moving it.

Range of motion: Keep your spine straight and stiff throughout the movement. This is a single-leg hip-hinge movement; therefore the movement should be directed from your hip region. The abdominal and latissimus dorsi muscles will help to stabilize and straighten the spine while the gluteus maximus and hamstrings on the same side as your stance leg generate hip extension. Tighten the gluteal muscles on the same side as your standing leg to stand straight up.

VARIATION

Floor Single-Leg Deadlift With Double Kettlebells

This exercise started from the top, after you deadlifted the kettlebells with both legs. You can start from the bottom as well. Move your stance and free leg into position and reach for the kettlebells, keeping your spine in proper alignment. Deadlift the kettlebells up and then set the kettlebells back on the ground, either tapping them or resting them for a short time.

3

SWING

Explosive. Tension. Hips. These are words that are well suited to describe the kettlebell swing. The swing is the dynamic brother of the grind deadlift. Both movements work muscles in similar ways, but the kettlebell swing is a unique feature of kettlebell training. Mark Reifkind, a StrongFirst master instructor and former powerlifter, is quoted as saying, "The swing is the center of the kettlebell universe."

What is the swing? Does it really help you build strength in all the right muscles? Will performing the swing make you fitter, able to run faster and jump higher, a better athlete, or have better posture? The answer to all these questions is a resounding "Yes!"

The swing, while strongly working the primary and secondary muscles, tends to work the entire body. Your cardiovascular system benefits; consistently performing the swing can lower your heart rate and blood pressure. Your nervous system will improve its coordination and ability to maintain proper posture while moving the kettlebell, whether in a grind exercise or with the dynamic swing. Your muscular system will increase in strength and resiliency.

The swing can be done with one kettlebell, two kettlebells, inside the legs, outside the legs, with one hand, or with two hands. The basic idea of starting in a hip-hinge position and standing up strongly applies to each variation.

One of the tenets in learning how to perform the swing is that this is a hip-generated movement. It's not a lumbar or a knee movement, but a hip movement. You could say the spine and the knee come along for the ride. As part of that process, they also become a lot stronger and move better.

When performing the kettlebell swing inside the legs, I teach my students to keep the kettlebell near the top of the small triangle on the descent of the swing or at the top of the hike pass. What is the small triangle? When you are looking at someone as they are standing, a large triangle is formed by the floor and the legs. The small triangle is still utilizing the legs, but the bottom of the triangle is at your knee level (figure 3.1). The small triangle is referred to throughout this book. Keeping the kettlebell high (near the top of the small triangle) decreases potential lower back discomfort and also improves the use of the stretch reflex during the swing, further enhancing the neuromuscular benefits achieved with the swing.

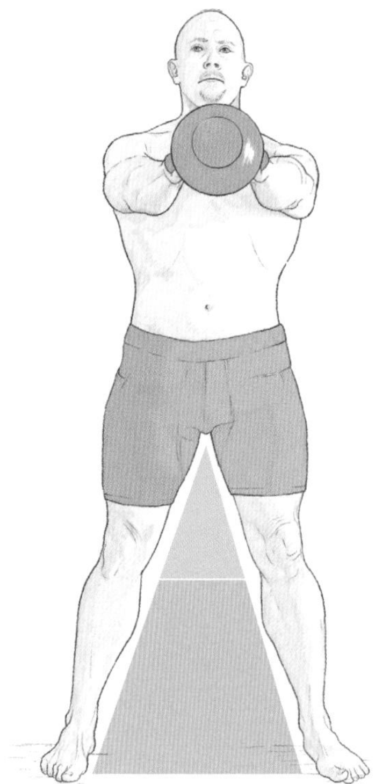

FIGURE 3.1 Small and large triangles in the kettlebell swing inside the legs.

Some of the potential benefits of performing the swing are that you will be able to jump higher, deadlift more, run faster, and have better posture. (See the volleyball player and sprinter in figure 3.2.) This is all from doing one exercise! I knew one kettlebell coach who successfully used the swing and the get-up, a dynamic paired with a grind, as the only exercises for his high school American football players! This is a perfect pairing of a ballistic and a grind.

Adding to what you learned from the kettlebell deadlifts, making sure that you are linked up during all phases of the swing will help you with exercises throughout this book. What does *linked up* mean? For starters, when you are at the top of the swing, your feet are rooted, your kneecaps are pulled up, your gluteal muscles are squeezed, your abdominal and latissimus dorsi muscles are tight and strongly contracted, and your shoulders are packed. Your head is on top of your shoulders, looking straight ahead.

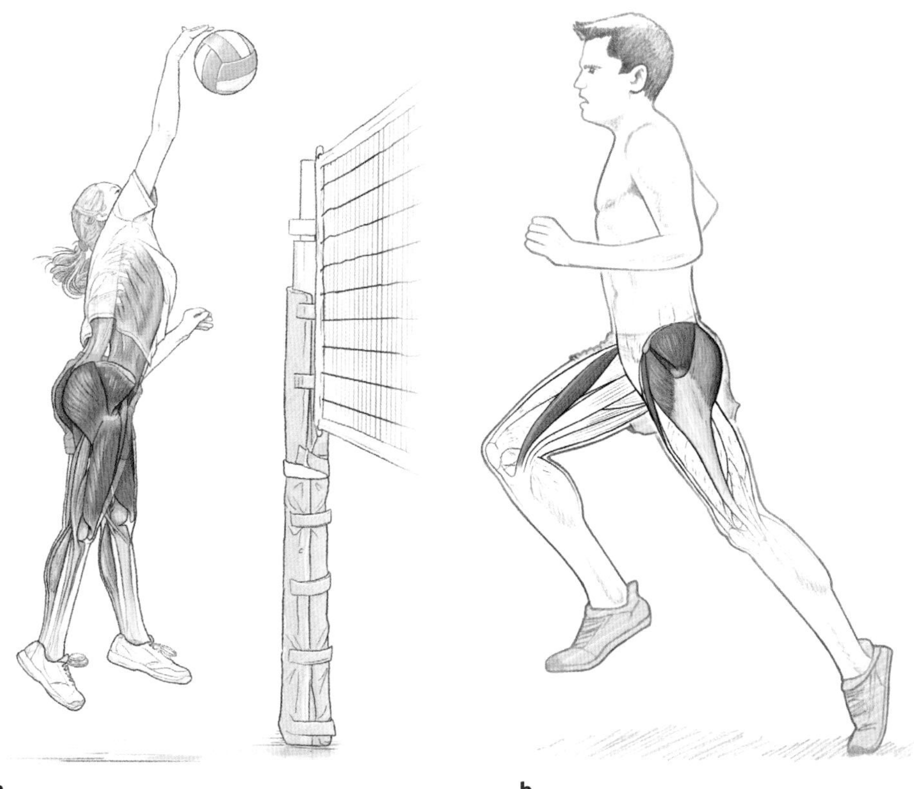

FIGURE 3.2 The kettlebell swing benefits athletes in many sports, including *(a)* volleyball and *(b)* sprinting.

Saying that your shoulders are packed means standing up and placing your arms parallel to the floor. Without moving anything else but your arms, reach for something in front of you, allowing yourself to protract your upper back. If you had a kettlebell in your hand and you were swinging, you would be leaking strength at this time. Now pull your shoulders back and down. You are now linked up and not leaking strength. This position is ideal at the top of the swing.

A common coaching proverb is, "The difference between an average athlete and an elite athlete is that the elite athlete can perform the basics much better." The kettlebell swing is a basic but powerful movement. The carryover to overall health and athleticism is huge and substantial.

As my friend and fellow kettlebell enthusiast Fabio Zonin said, "I performed a perfect swing . . . once." Keep practicing and training the swing regularly and you will see the tremendous benefits that it will give you.

EXERCISE FOCUS FOR KETTLEBELL SWINGS

The exercises in this chapter can be organized into two categories:

1. Kettlebell swing, inside legs
2. Kettlebell swing, outside legs

Each category of the kettlebell swing provides different ways to help you learn proper technique and develop strength. Here, we take a closer look at the focus of the exercises in each category.

Kettlebell Swing, Inside Legs

The kettlebell two-hand swing is a powerful movement that utilizes the body positioning that an athlete learns from performing the kettlebell deadlift in the previous chapter. Some of the benefits of performing the kettlebell swing include developing a strong posterior chain, which is used in daily activities and many sports to help you run and jump; creating and holding a vertical plank position momentarily at the top of the lift; learning proper tension-developing skills, which will be used later on with other strength movements; and learning the proper setup for the single-hand kettlebell swing and other related movements.

A progression of the kettlebell two-hand swing is the kettlebell one-hand swing. In addition to the aforementioned benefits, you will learn to keep your shoulders and hips squared up throughout this movement even though you are using an asymmetrical load, to be able to recruit additional muscles versus the two-hand swing: the contralateral gluteus medius and minimus and quadratus lumborum will be further activated during this swing, and you will learn the proper setup for the alternating single-hand kettlebell swing. In addition to the advantages of performing the one-hand swing, the alternating one-hand swing teaches you to strengthen your hip snap at the top of the vertical plank, thereby increasing the float time of the kettlebell, allowing you to quickly switch hands for the next repetition. Finally, the double kettlebell swing further enhances holding the vertical plank position at the top of the lift while at the same time further strengthening how your hips extend with heavier weight.

Kettlebell Swing, Outside Legs

When performing the kettlebell swing with the kettlebell outside your legs, having attention to detail is important regarding your body positioning prior to starting the exercise. Similar to swinging the kettlebell with one hand inside the legs, performing it outside the legs will also develop a strong posterior chain, which is used in daily activities in many sports to help you run and jump. It will also help you to keep your shoulders and hips squared up throughout this movement, even though you're using an asymmetrical load. Since the kettlebell is further laterally from your center of mass, this will make your body work harder to maintain a straight spine throughout the movement.

Some other benefits will be creating and holding a vertical plank position momentarily at the top of the lift and learning the tension-developing skills that are needed when doing asymmetrical exercises. This would benefit many sports where there is not bilateral movement of the body, such as swinging a baseball bat, throwing a ball, or playing golf. The double kettlebell swing outside your legs will teach you to extend your hips with heavier weight and to keep your shoulders and hips squared up while performing this exercise.

SINGLE KETTLEBELL TWO-HAND SWING, INSIDE LEGS

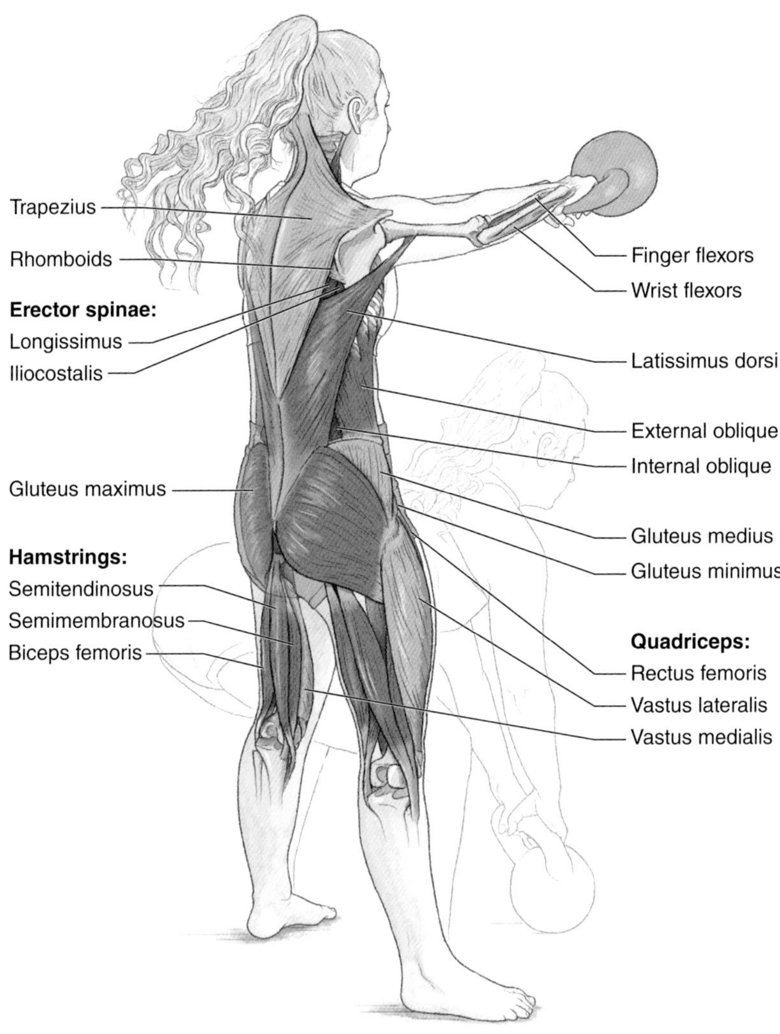

Execution

1. Place your feet about a foot (0.3 m) behind the kettlebell. Take a shoulder-width stance behind the kettlebell with your toes pointed slightly outward.
2. Keeping the hips above the knees and below the shoulders, grip the kettlebell handle with both hands, placing your body into a hip-hinge position. You should feel tension in your hamstrings. Tilt the handle toward you, allowing it to become an extension of your arms.
3. Grip the ground with your toes; tighten your latissimi dorsi, abdominal muscles, and your grip; and hike pass the kettlebell back between your legs without changing your hip-hinge position. Target the kettlebell to the small triangle above your knees.
4. Stand up with the kettlebell until you are fully erect, squeezing your gluteal muscles hard, pulling your kneecaps up strongly, and bracing your abdominal muscles hard. The kettlebell will swing up to approximately your chest level.
5. Once the kettlebell starts descending, use your latissimus dorsi muscles to pull it down. Keep your vertical plank position until the last moment, and then suddenly perform a hip hinge, continuing to bring the kettlebell between your legs near the top of the small triangle. (Think of this movement as getting out of the way of the kettlebell.) Doing this will keep the kettlebell in the small triangle between your legs during your hip hinge.
6. This is a ballistic exercise. Immediately go into the next repetition.

Muscles Involved

Primary: Erector spinae (iliocostalis, longissimus, spinalis), gluteus maximus, hamstrings (semitendinosus, semimembranosus, biceps femoris), latissimus dorsi, rectus abdominis, external oblique, internal oblique, transversus abdominis

Secondary: Trapezius, rhomboids, quadriceps (rectus femoris, vastus lateralis, vastus medialis, vastus intermedius), gluteus medius, gluteus minimus, forearms (wrist flexors, finger flexors)

(continued)

SINGLE KETTLEBELL TWO-HAND SWING, INSIDE LEGS *(continued)*

Anatomic Focus

Hand spacing: Place both hands next to each other on the handle of the kettlebell.

Grip: Use a double overhand grip (pronate both hands). Try to break the handle, thereby strongly activating the latissimus dorsi.

Stance: Position the legs approximately one foot (0.3 m) behind the kettlebell and shoulder-width apart. Angle the toes slightly outward.

Trajectory: Keep your elbows straight on the upswing and the handle of the kettlebell horizontal. The swing is a horizontal projection of force at the top.

Range of motion: Keeping your arms straight, hike the kettlebell back between your legs, aiming for the small triangle above your knees. Pretend that you are playing American football and are snapping the ball to a friend 10 to 15 yards (9-14 m) behind you. Keep your hip-hinge position the same while hiking the kettlebell. When you have reached the end of the hike, stand up quickly. Your toes will be rooted to the ground, your kneecaps pulled up, your gluteal muscles squeezed, and your abdominal and latissimus dorsi muscles strongly engaged. As in the kettlebell deadlift, the erector spinae, abdominal muscles, and latissimus dorsi muscles will help to stabilize and straighten the spine while the gluteus maximus and hamstrings generate hip extension. Your spine should be straight and stiff throughout the movement. Antishrug your shoulders, further activating your latissimi dorsi. Do not overextend your spine.

VARIATION

Dead Stop Swing (Power Swing)

This variation is performed as detailed in the single kettlebell two-hand swing, inside legs except that every rep starts and stops on the ground. This exercise is very powerful and explosive and is generally done for three to five repetitions per set. It can be performed with either one or two hands on a kettlebell. It is better performed inside your legs. It can be used to improve your hike back between your legs at the start or as a standalone exercise.

SINGLE KETTLEBELL ONE-HAND SWING, INSIDE LEGS

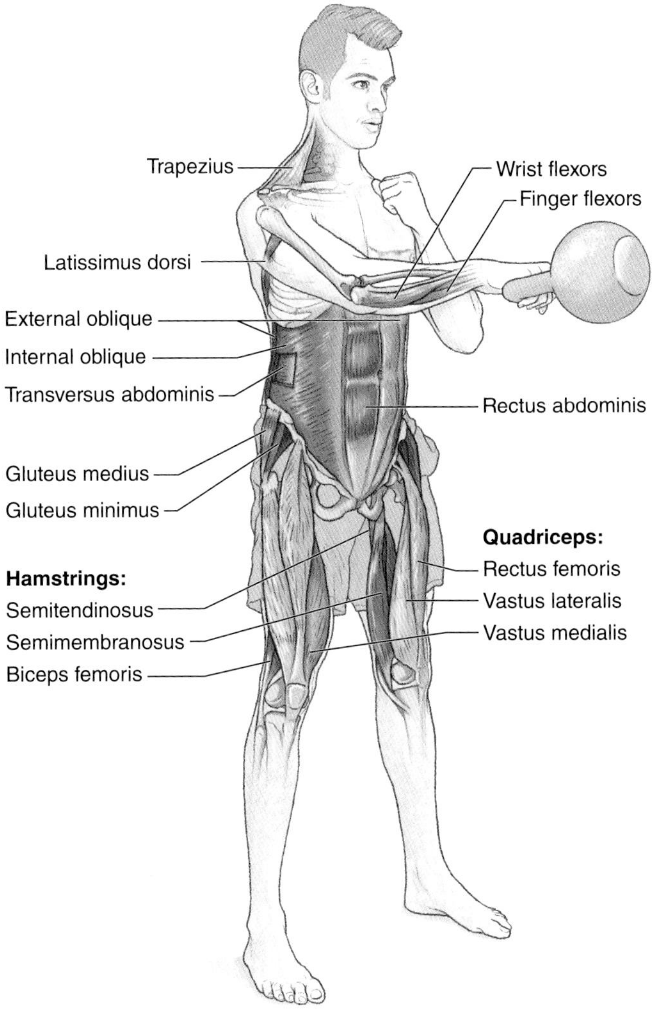

Swing inside legs.

Execution

1. Place your feet about a foot (0.3 m) behind the kettlebell. Take a shoulder-width stance behind the kettlebell with your toes pointed slightly outward.
2. Keeping the hips above the knees and below the shoulders, grip the kettlebell handle with one hand, placing your body into a hip-hinge position. You should feel tension in your hamstrings. Tilt the handle toward you, allowing it to become an extension of your arm.

(continued)

SINGLE KETTLEBELL ONE-HAND SWING, INSIDE LEGS *(continued)*

3. Place your nonworking hand beside the kettlebell but not on it. Doing this will help keep your shoulders square at the beginning.
4. Grip the ground with your toes; tighten the latissimus dorsi and abdominal muscles, and grip; and hike pass the kettlebell back between your legs without changing your hip-hinge position. Target the kettlebell to the small triangle above your knees.
5. Stand up with the kettlebell until you are fully erect, squeezing your gluteal muscles hard, pulling your kneecaps up strongly, and bracing your abdominal muscles hard. The kettlebell will swing up to approximately your chest level.
6. At the top, your shoulders and hips should be squared up, even though only one hand is on the kettlebell.
7. Once the kettlebell starts descending, use your latissimus muscle to pull it down. Keep your vertical plank position until the last moment, and then get out of the way of the kettlebell. Doing this will keep the kettlebell in the small triangle between your legs during your hip hinge.
8. Complete all repetitions on one side, then switch hands.

Muscles Involved

Primary: Erector spinae (iliocostalis, longissimus, spinalis), gluteus maximus, gluteus medius, gluteus minimus, quadratus lumborum, hamstrings (semitendinosus, semimembranosus, biceps femoris), latissimus dorsi, rectus abdominis, external oblique, internal oblique, transversus abdominis

Secondary: Trapezius, rhomboids, quadriceps (rectus femoris, vastus lateralis, vastus medialis, vastus intermedius), forearms (wrist flexors, finger flexors)

Anatomic Focus

Hand spacing: Place one hand in contact with the middle of the kettlebell handle, clasping it loosely with your fingers.

Grip: Use a single overhand grip (pronate the hand).

Stance: Position the legs approximately one foot (0.3 m) behind the kettlebell and shoulder-width apart. Point the toes slightly outward.

Trajectory: Keep your elbow straight on the upswing and the handle of the kettlebell horizontal. The swing is a horizontal projection of force at the top.

Range of motion: Keeping your arm straight, hike the kettlebell back between your legs, aiming for the small triangle above your knees. Pretend that you are snapping a ball to a friend 10 to 15 yards (9-14 m) behind you as in American football. Keep your hip-hinge position the same while hiking the kettlebell. When you have reached the end of the hike, stand up quickly. Your toes will be rooted to the ground, your kneecaps pulled up, your gluteal muscles squeezed, and your abdominal and latissimus dorsi muscle strongly engaged. As in the kettlebell deadlift, the erector spinae, abdominal muscles, and latissimus dorsi muscles will help to stabilize and straighten the spine while the gluteus maximus and hamstrings generate hip extension. Your spine should be straight and stiff throughout the movement. Antishrug your shoulders, further activating your latissimi dorsi. Do not overextend your spine.

My recommendation is to have your nonworking arm mimic what the working arm is doing during the one-handed swing. If you wish, when the kettlebell is at the top of the movement, you can place your nonkettlebell arm in a guard position instead of having it out in front. Making sure that the nonworking arm is moving in this way will help avoid rotation during the exercise.

Technique note: Switch your starting hand at the beginning of each succeeding set. Doing this prevents having a favorite start hand and will help in keeping a balanced body while training.

SINGLE KETTLEBELL ONE-HAND SWING, ALTERNATING HANDS, INSIDE LEGS

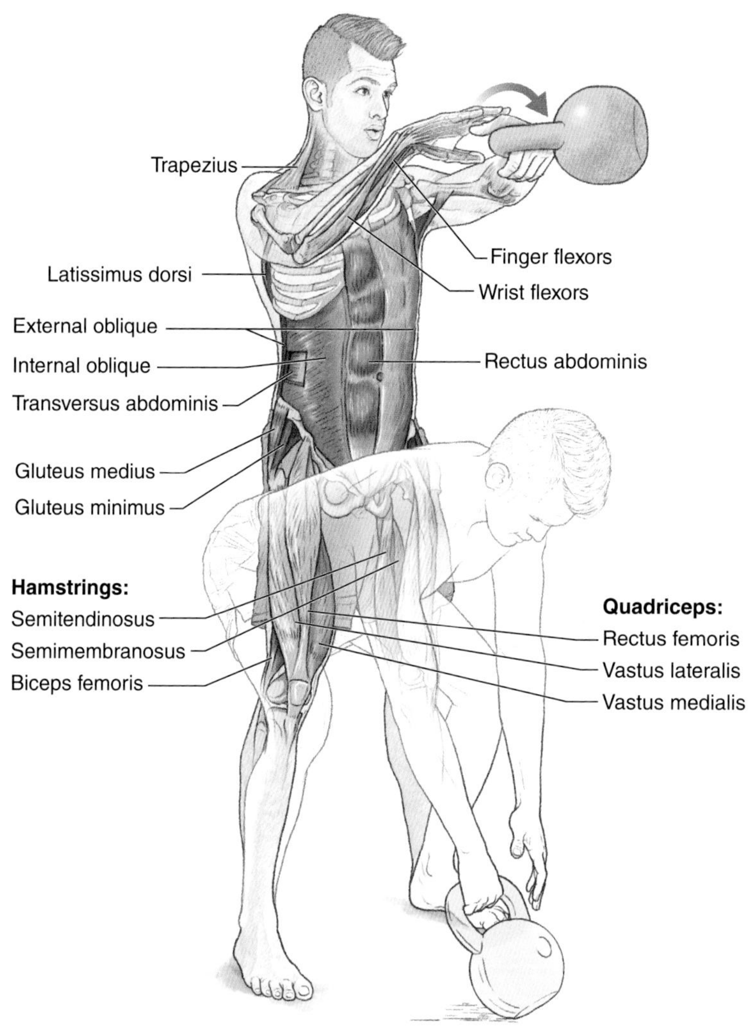

Execution

1. Place your feet about a foot (0.3 m) behind the kettlebell. Take a shoulder-width stance behind the kettlebell with your toes pointed slightly outward.
2. Keeping the hips above the knees and below the shoulders, grip the kettlebell handle with one hand, placing your body into a hip-hinge position. You should feel tension in your hamstrings. Tilt the handle toward you, allowing it to become an extension of your arm.
3. Place your nonworking hand beside the kettlebell but not on it. Doing this will help keep your shoulders square at the beginning.
4. Grip the ground with your toes; tighten the latissimi dorsi, abdominal muscles, and grip; and hike pass the kettlebell back between your legs without changing your hip-hinge position. Target the kettlebell to the small triangle above your knees.
5. Stand up with the kettlebell until you are fully erect, squeezing your gluteal muscles hard, pulling your kneecaps up strongly, and bracing your abdominal muscles hard. The kettlebell will swing up to approximately your chest level.
6. At the top, your shoulders and hips should be squared up, even though only one hand is on the kettlebell.
7. Once you have achieved the vertical plank with a strong glute contraction, allow the kettlebell to momentarily float in the air while at the same time switching hands on the handle.
8. Once the kettlebell starts descending, use your latissimus muscle to pull it down. Keep your vertical plank position until the last moment, and then get out of the way of the kettlebell. Doing this will keep the kettlebell in the small triangle between your legs during your hip hinge.

Muscles Involved

Primary: Erector spinae (iliocostalis, longissimus, spinalis), gluteus maximus, gluteus medius, gluteus minimus, quadratus lumborum, hamstrings (semitendinosus, semimembranosus, biceps femoris), latissimus dorsi, rectus abdominis, external oblique, internal oblique, transversus abdominis

Secondary: Trapezius, rhomboids, quadriceps (rectus femoris, vastus lateralis, vastus medialis, vastus intermedius), forearms (wrist flexors, finger flexors)

(continued)

SINGLE KETTLEBELL ONE-HAND SWING, ALTERNATING HANDS, INSIDE LEGS *(continued)*

Anatomic Focus

Hand spacing: Place one hand in contact with the middle of the kettlebell handle. Clasp it loosely with your fingers. Switch the hand on the kettlebell at the top of the swing.

Grip: Use a single overhand grip (pronate the hand).

Stance: Position the legs approximately one foot (0.3 m) behind the kettlebell and shoulder-width apart. Point your toes slightly outward.

Trajectory: Keep your elbow straight on the upswing and the handle of the kettlebell horizontal. The swing is a horizontal projection of force at the top.

Range of motion: Keeping your arm straight, hike the kettlebell back between your legs, aiming for the small triangle above your knees. Pretend that you are snapping a ball to a friend 10 to 15 yards (9-14 m) behind you as in American football. Keep your hip-hinge position the same while hiking the kettlebell. When you have reached the end of the hike stand up quickly. Your toes will be rooted to the ground, your kneecaps pulled up, your gluteal muscles squeezed, and your abdominal and latissimus dorsi muscles strongly engaged. As in the kettlebell deadlift, the erector spinae, abdominal muscles, and latissimi dorsi will help to stabilize and straighten the spine while the gluteus maximus and hamstrings generate hip extension. Your spine should be straight and stiff throughout the movement. Antishrug your shoulders, further activating your latissimi dorsi. Do not overextend your spine.

My recommendation is to have your nonworking arm mimic what your working arm is doing during the one-hand swing. Making sure that the nonworking arm is moving in this way will help avoid rotation during the exercise and be ready to accept the switch of the hands at the top of the movement.

Technique note: Switch your start hand at the beginning of each succeeding set. Doing this prevents having a favorite start hand and will help keeping a balanced body while training.

DOUBLE KETTLEBELL SWING, INSIDE LEGS

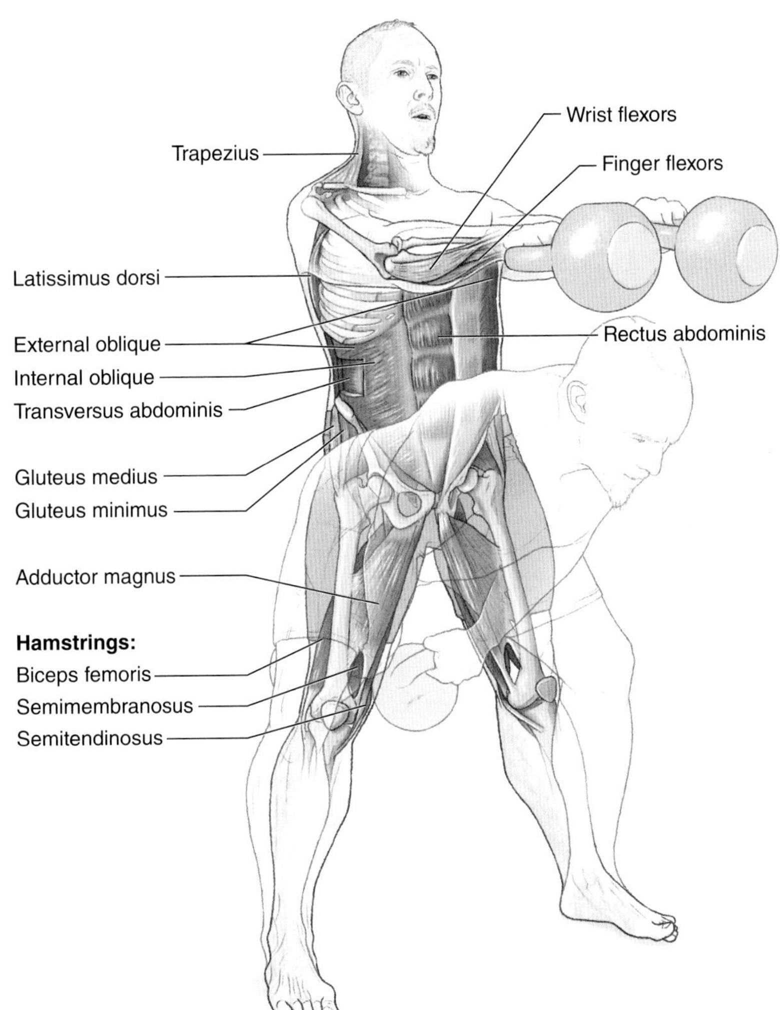

Execution

1. Put two kettlebells next to each other with their handles in line.
2. Place your feet about a foot (0.3 m) behind the kettlebells. Take a stance wider than shoulder width (wide enough so you won't hit your knees) behind the kettlebells with your toes pointed slightly outward.
3. Keeping the hips above the knees and below the shoulders, grip each kettlebell handle with a hand, placing your body into a hip-hinge position. You should feel tension in your hamstrings. Tilt the handles toward you, allowing them to become an extension of your arms.

(continued)

DOUBLE KETTLEBELL SWING, INSIDE LEGS *(continued)*

4. Grip the ground with your toes; tighten the latissimi dorsi, abdominal muscles, and grip; and hike pass the kettlebells back between your legs without changing your hip-hinge position. Target the kettlebells to the small triangle above your knees.
5. Stand up with the kettlebells until you are fully erect, squeezing your gluteal muscles hard, pulling your kneecaps up strongly, and bracing your abdominal muscles hard. The kettlebells will swing up to approximately your stomach to chest level.
6. Once the kettlebells start descending, use your latissimi dorsi to pull them down. Keep your vertical plank position until the last moment, and then get out of the way of the kettlebells. Doing this will keep the kettlebells in the small triangle between your legs during your hip hinge.

Muscles Involved

Primary: Erector spinae (iliocostalis, longissimus, spinalis), gluteus maximus, hamstrings (semitendinosus, semimembranosus, biceps femoris), adductor magnus, latissimus dorsi, rectus abdominis, external oblique, internal oblique, transversus abdominis

Secondary: Trapezius, rhomboids, quadriceps (rectus femoris, vastus lateralis, vastus medialis, vastus intermedius), gluteus medius, gluteus minimus, forearms (wrist flexors, finger flexors)

Anatomic Focus

Hand spacing: Place each hand on its own kettlebell.

Grip: Use a double overhand grip (pronate both hands).

Stance: Position the legs approximately one foot (0.3 m) behind the kettlebell and more than shoulder-width apart. Angle the toes slightly outward.

Trajectory: Keep your elbows straight on the upswing and the handle of the kettlebell horizontal. The swing is a horizontal projection of force at the top.

Range of motion: Keeping your arms straight, hike the kettlebells back between your legs, aiming for the small triangle above your knees. Pretend that you are snapping a ball to a friend 10 to 15 yards (9-14 m) behind you as in American football. Keep your hip-hinge position the same while hiking the kettlebells. When you have reached the end of the hike, stand up quickly. Your toes will be rooted to the ground, your kneecaps pulled up, your gluteal muscles squeezed, and your abdominal muscles and latissimi dorsi strongly engaged. As in the kettlebell deadlift, the erector spinae, abdominal muscles, and latissimi dorsi will help to stabilize and straighten the spine while the gluteus maximus and hamstrings generate hip extension. Your spine should be straight and stiff throughout the movement. Antishrug your shoulders, further activating your latissimi dorsi. Do not overextend your spine.

Technique note: With the double kettlebells, you will be swinging a heavier load with a wider stance. One thing of note is that the kettlebells could throw you off balance, so you may need to lean back from your ankles, not your lower back, at the top of the swing. However, during the hike pass, you will need to shift your weight back to the center of your feet. Additionally, depending on how heavy the kettlebells are, the height of the kettlebells at the top of the swing could be anywhere from your chest height down to your waist area.

SINGLE KETTLEBELL ONE-HAND SWING, OUTSIDE LEGS

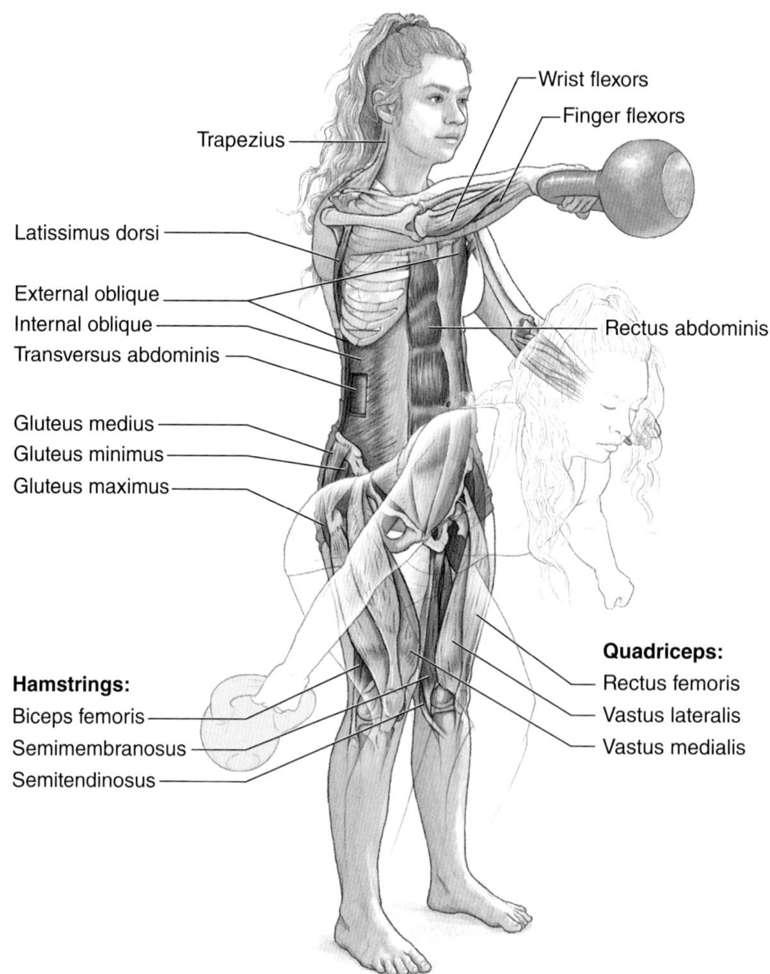

Execution

1. Stand with your feet hip-width apart or less. Have the kettlebell just outside one foot and your feet just behind it.
2. Keeping the hips above the knees and below the shoulders, grip the kettlebell handle with one hand, placing your body into a hip-hinge position. You should feel tension in your hamstrings. Tilt the handle toward you, allowing it to become an extension of your arm.
3. Keep your other hand to the side of your body in a position as if it was lifting a kettlebell. Doing this will help keep your shoulders square at the beginning.
4. Grip the ground with your toes; tighten the latissimi dorsi, abdominal muscles, and grip; and hike pass the kettlebell back outside your leg without changing your hip-hinge position.
5. Stand up with the kettlebell until you are fully erect, squeezing your gluteal muscles hard, pulling your kneecaps up strongly, and bracing your abdominal muscles hard. The kettlebell will swing up to approximately your stomach or chest level.
6. At the top, your shoulders and hips should be squared up, even though only one hand is on the kettlebell.
7. Once the kettlebell starts descending, use your latissimus dorsi muscle to pull it down. Keep your vertical plank position until the last moment, then hip hinge at the same time as the kettlebell goes back.
8. Complete the repetitions on one side, then switch hands.

Muscles Involved

Primary: Erector spinae (iliocostalis, longissimus, spinalis), gluteus maximus, gluteus medius, gluteus minimus, quadratus lumborum, hamstrings (semitendinosus, semimembranosus, biceps femoris), latissimus dorsi, rectus abdominis, external oblique, internal oblique, transversus abdominis

Secondary: Trapezius, rhomboids, quadriceps (rectus femoris, vastus lateralis, vastus medialis, vastus intermedius), forearms (wrist flexors, finger flexors)

(continued)

SINGLE KETTLEBELL ONE-HAND SWING, OUTSIDE LEGS *(continued)*

Anatomic Focus

Hand spacing: Place one hand in contact with the middle of the kettlebell handle. Clasp it loosely with your fingers.

Grip: Use a single overhand grip (pronate the hand).

Stance: Position the feet just behind the kettlebell and hip-width or less apart. Point the toes slightly outward.

Trajectory: Keep your elbow straight on the upswing and the handle of the kettlebell horizontal. The swing is a horizontal projection of force at the top.

Range of motion: Keeping your arm straight, hike the kettlebell back beside your leg. Using what you learned in chapter 2 from the single kettlebell deadlift with one hand, outside legs (also known as the suitcase deadlift), turn that grind exercise into a dynamic one. Initially the height of the kettlebell will be at your waist level, but as you continue swinging it, the height will get higher, possibly to your chest. Since the kettlebell is outside your leg, the relative work on your nonworking arm side will be greater as the load is further from your center, when compared to the single kettlebell one-hand swing, inside legs. Make sure your shoulders and hips are squared up at both the bottom and the top of the swing. When you have reached the end of the hike, stand up quickly. Your toes will be rooted to the ground, your kneecaps pulled up, your gluteal muscles squeezed, and your abdominal muscles and latissimi dorsi strongly engaged.

As in the suitcase deadlift, the erector spinae, abdominal muscles, and latissimi dorsi will help to stabilize and straighten the spine while the gluteus maximus and hamstrings generate hip extension. Your spine should be straight and stiff throughout the movement. Antishrug your shoulders, further activating your latissimi dorsi. Do not overextend your spine.

I recommend you have your nonworking arm mimic the working arm during the one-hand swing. If you wish, when the kettlebell is at the top of the movement, you can place the nonkettlebell arm in a guard position instead of having it out in front. Making sure that the nonworking arm is moving in this way will help avoid rotation during the exercise.

DOUBLE KETTLEBELL SWING, OUTSIDE LEGS

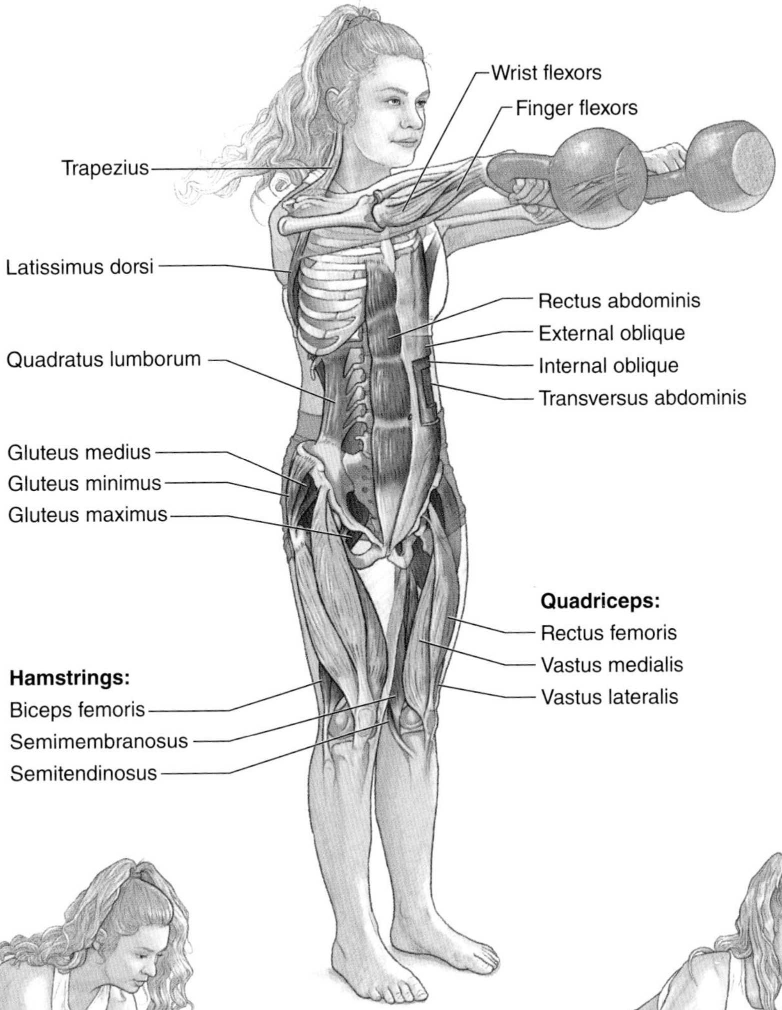

Wrist flexors
Finger flexors
Trapezius
Latissimus dorsi
Rectus abdominis
External oblique
Internal oblique
Transversus abdominis
Quadratus lumborum
Gluteus medius
Gluteus minimus
Gluteus maximus

Quadriceps:
Rectus femoris
Vastus medialis
Vastus lateralis

Hamstrings:
Biceps femoris
Semimembranosus
Semitendinosus

Start position.

Hike pass.

(continued)

DOUBLE KETTLEBELL SWING, OUTSIDE LEGS *(continued)*

Execution

1. Stand with your feet hip-width apart or less. Have the kettlebells just outside your feet and slightly in front of them.
2. Keeping the hips above the knees and below the shoulders, grip a kettlebell handle in each hand, placing your body into a hip-hinge position. You should feel tension in your hamstrings. Tilt the handle toward you, allowing it to become an extension of your arm.
3. Grip the ground with your toes; tighten the latissimi dorsi, abdominal muscles, and grip; and hike pass the kettlebell back outside your legs without changing your hip-hinge position.
4. Stand up with the kettlebells until you are fully erect, squeezing your gluteal muscles hard, pulling your kneecaps up strongly, and bracing your abdominal muscles hard. The kettlebells will swing up approximately to the level of your stomach or chest.
5. At the top, your shoulders and hips should be squared up.
6. Once the kettlebells start descending, use your latissimus muscle to pull them down. Keep your vertical plank position until the last moment, then hip hinge at the same time as the kettlebells move back.

Muscles Involved

Primary: Erector spinae (iliocostalis, longissimus, spinalis), quadratus lumborum, hamstrings (semitendinosus, semimembranosus, biceps femoris), latissimus dorsi, rectus abdominis, external oblique, internal oblique, transversus abdominis

Secondary: Trapezius, rhomboids, quadriceps (rectus femoris, vastus lateralis, vastus medialis, vastus intermedius), gluteus medius, gluteus minimus, forearms (wrist flexors, finger flexors)

Anatomic Focus

Hand spacing: Place each hand in contact with the middle of the kettlebell handle, clasping it loosely with your fingers.

Grip: Use a double overhand grip (pronate both hands).

Stance: Position your feet just behind the kettlebell and hip-width or less apart. Point the toes slightly outward.

Trajectory: Keep your elbows straight on the upswing and the handles of the kettlebell horizontal. The swing is a horizontal projection of force at the top.

Range of motion: Keeping your arms straight, hike the kettlebells back beside your legs. Using what you learned in chapter 2 from the double kettlebell deadlift, outside legs, turn that grind exercise into a dynamic one. Initially the height of the kettlebells will be at your waist level, but as you continue swinging them, the height will get higher, possibly to your chest. Since the kettlebells are outside your legs, the relative work on your gluteus maximus could be potentially greater as the load is farther from your center.

Make sure your shoulders and hips are squared up at both the bottom and the top of the swing. When you have reached the end of the hike, stand up quickly. Your toes will be rooted to the ground, your kneecaps pulled up, your gluteal muscles squeezed, and your abdominal muscles and latissimi dorsi strongly engaged. Your erector spinae, abdominal muscles, and latissimi dorsi will help to stabilize and straighten the spine while the gluteus maximus and hamstrings generate hip extension. Your spine should be straight and stiff throughout the movement. Antishrug your shoulders, further activating your latissimi dorsi. Do not overextend your spine.

Technique note: With the double kettlebells, you will be swinging a heavier load with a narrower stance. One thing of note is that they could throw you off balance, so you may need to lean back from your ankles, not your lower back, at the top of the swing. However, during the hike pass, you will need to shift your weight back to the center of your feet. Additionally, depending on how heavy they are, the height of the kettlebells at the top of the swing could be anywhere from your chest height down to your waist area.

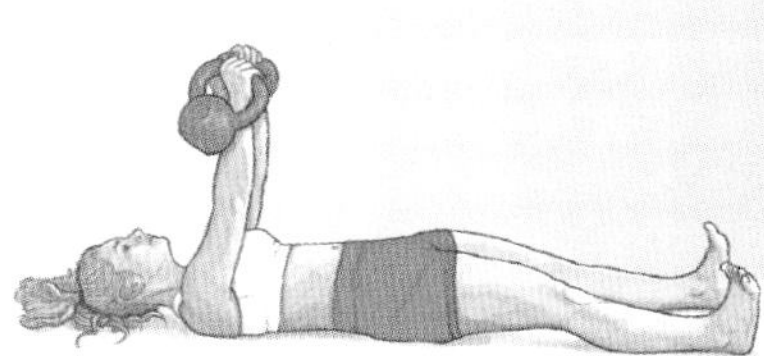

4

CLEAN AND PRESS

This chapter will discuss two kettlebell movements: the clean and the press. Although they are related movements, they are separate exercises. The relationship is that to perform a kettlebell press, you use a kettlebell clean to get into position. Without the kettlebell clean, there would be no kettlebell press, especially when pressing heavier kettlebells.

KETTLEBELL CLEAN

Intense words that describe the kettlebell clean include *speed, tension, fluid motion, power,* and *vertical projection of force.* Watching it in real time is like listening to classical music with heavy metal mixed in. First you have the swing part of the clean: going from the floor, hiking it back between your legs and standing up, forcefully squeezing your gluteal and abdominal muscles. Unlike the swing, which is a horizontal projection of force, the end stage of the clean changes this. With your upper arm staying against your side, the force of the bell is going vertically, not horizontally. At the last moment, the kettlebell rolls around your wrist and finishes in the rack position. This is where the music goes from classical to heavy metal. You are planking but in a vertical, standing position. Everything is squeezed and tightened.

Remember, your clean is only as good as your kettlebell swing. Work on your swing before practicing the clean.

A kettlebell clean is not an Olympic weightlifting–style clean. Your shoulders stay down and you do not drop under the kettlebell. Instead of your elbows coming up, they stay in and down.

Although cleans can be a stand-alone exercise, they usually are combined with other movements. The clean is used as a means to an end. For example, to be able to perform a kettlebell front squat, you need to perform a kettlebell clean to the rack position first. This is also true with the press, rack carries (discussed in chapter 9), and various other kettlebell exercises.

You can also do complexes with cleans. Performing a clean plus press plus squat is one example of an exercise complex (i.e., combining multiple exercises into one set) that takes three kettlebell exercises and combines them into a giant exercise that brings about many benefits, including strength endurance and mental fortitude.

KETTLEBELL PRESS

In a kettlebell press, you use strength, explode upward, push, breathe, and exert lots of tension. The press is a grind movement—an almost pure strength move. I have always been good at pressing, so learning the kettlebell press came easy to me. There are others who dislike the grind aspect of the press or who don't like the difficulty of it. My comment to them is eventually they will learn to like it and the improvements to their upper-body development.

As you will see in this chapter, the press has several variations. My recommendation is to learn the basic press movement first and move on to the variations. The kettlebell get-up plus press, which was already discussed in chapter 5, is an example of a variation.

From a physics standpoint, why I love the kettlebell press is what it can do for my gluteal and abdominal muscles, in addition to my deltoids, triceps, pectorals, and latissimi dorsi. Here is what I mean: After you perform a clean to bring the kettlebell to the rack position, using, let's say, a 24-kilogram (53 lb) kettlebell, your body weight just gained that additional weight. If the typical center of mass of a human is around the waist area, after performing a kettlebell clean, the center of mass just moved higher than that. This is due to the weight of the kettlebell being above the center of mass. As you start pressing the weight up, that center of mass also goes up and continues going higher until you stop pressing. What this does to your gluteal and abdominal muscles is put an extra strain on them to keep their same position as when the kettlebell was in the rack position. If they don't keep this position, the press will end up extending the lumbar spine and flaring your rib cage under load, which isn't healthy.

BOTTOM-UP CLEAN AND PRESS

The last two exercises use the bottom-up position. *Bottom-up* means the body of the kettlebell is above the handle, not below it. These two movements will elevate your ability to increase your total-body tension tenfold.

First I'll discuss the bottom-up clean, both single and double kettlebell. Once you get the kettlebell into the rack position, freeze. Stay there for a few seconds. Tense every muscle in your body. Try to crush the handle. Remember everything that I have mentioned previously about tensing to this point (i.e., rooting your feet, pulling your kneecaps up, squeezing your gluteal muscles, etc.) and increase the tension even further.

Second, I'll discuss the bottom-up press. Again, single and double kettlebells apply here. This takes the tension meter and goes off the charts. Slowly press the kettlebell up and pull it down with the same speed.

I would recommend spending some quality time with the single kettlebell bottom-up versions of the clean and the press before venturing into the double kettlebell versions. Using these bottom-up variations will benefit you tremendously. You will also develop an indomitable grip on the kettlebell this way. You will thank me later.

Learning to perform the clean and the press will add many benefits to your body and also increase your performance in sports and make you more resilient in the event of an injury.

EXERCISE FOCUS FOR KETTLEBELL CLEANS AND PRESSES

Cleans and presses provide different ways to help you learn proper technique and develop strength. Here, we take a closer look at the focus of the exercises in each category.

Cleans

In chapter 3, you learned to do a kettlebell swing, and here you will learn to effectively perform a kettlebell clean. The clean is an exercise by itself, but it is also a gateway to performing other exercises with the kettlebell, such as the front squat, the press, and the rack carry. Some of the advantages of performing the kettlebell clean include developing a strong posterior chain, which is used in daily activities such as shoveling snow and pulling a heavy suitcase, and sports, such as powerlifting, weightlifting, and playing American football (figure 4.1).

The rack-position part of the kettlebell clean is a fundamental posture in kettlebell training that serves as a transition point for several exercises, such as the kettlebell press and front squat. In the rack position, the kettlebell is held close to the body at chest level, with the following key points in mind:

1. Hand positioning: The kettlebell handle should be held diagonally across the palm, with the thumb pointed toward the body and the pinky finger away from you. The handle should rest between the base of the fingers and the heel of the palm.
2. Elbow placement: The elbow should be tucked close to the body, with the forearm held vertically. This helps to support the weight of the kettlebell on the bones of the forearm, reducing strain on the muscles.
3. Wrist alignment: The wrist should be kept straight and neutral, with no extension or flexion. This minimizes the risk of injury and ensures proper weight distribution.
4. Kettlebell position: The kettlebell should rest on the back side of the forearm and the upper part of the arm (biceps and shoulder).
5. Core engagement: Engage your abdominal and gluteal muscles strongly to maintain an upright torso while holding the kettlebell in the rack position.

Mastering the kettlebell rack position is important for safely and effectively performing several different kettlebell exercises.

While performing a single kettlebell clean, you are moving with an asymmetrical load and recruiting more muscles (the contralateral gluteus medius, gluteus minimus, and quadratus lumborum) than the double clean. In addition, you will be creating and holding a vertical plank position momentarily at the top of the lift isometrically, becoming one with the kettlebell and learning proper tension-developing skills. When performing a double kettlebell clean, you are using a heavier weight, which will require you to develop even more tension, especially at the rack position.

FIGURE 4.1 A defensive tackle in American football needs a strong posterior chain to break through a double team.

An advanced progression of the clean is the bottom-up clean, performed with a single kettlebell or double kettlebell. This further uses Sherrington's Law of Irradiation through the crush grip on the handle of the kettlebell throughout the movement, which will send neuromuscular signals up through your upper extremity, through your rotator cuff, and into various stabilizers in a stronger way then regular kettlebell cleans. Sherrington's Law of Irradiation states that a specific muscle that is working hard tends to recruit the muscles nearby, and if they are already part of the action taking place, it will further amplify the overall strength being professed. This will further develop your tension-developing skills, which will pay dividends while performing the press.

Presses

It is said that a good press is preceded by a good clean, which is preceded by a good swing. I have never seen a good press performed after a poor swing and clean. Similar to the kettlebell get-up's role in developing strong, resilient shoulders, the press also helps develop a strong upper body, particularly the arm and shoulders, which will be demonstrated effectively while performing martial arts, playing hockey, and playing golf (figure 4.2).

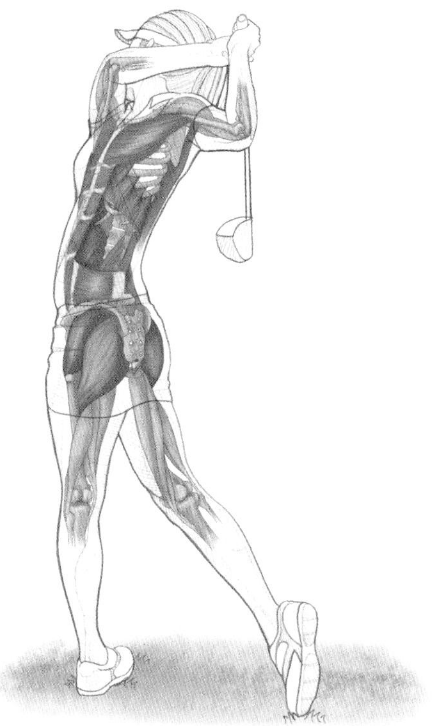

FIGURE 4.2 Presses develop strong arms and shoulders, perfect for a golfer.

Your overall center of mass is elevating during the press. Your gluteal and abdominal muscles need to be contracted strongly during both the single and double kettlebell press to prevent extending your lumbar spine and flaring your rib cage. A positive side effect of this is also a stronger press as it establishes a better foundation to press from. As mentioned during my discussion of the clean, when you are using single kettlebell, you are using an asymmetrical load, which will recruit additional muscles on the contralateral side during the movement. While performing a double kettlebell clean press, you will further recruit additional motor units as the double kettlebells rise above your head. Performing the single and double kettlebell bottom-up press will further enhance your ability to generate tension during this lift and in other high-tension sports such as strongman and track and field.

The Sots press, named after Russian weightlifter Viktor Sots, is an interesting movement in that while you are down in the bottom of a squat, you will be pressing the kettlebell overhead. This is an advanced progression that requires excellent mobility and strength in your feet, ankles, knees, hips, torso, and shoulders. Performing this exercise will also make you more resilient to injury and increase your performance.

The kettlebell floor press, both single and double, are very similar to the pressing of the kettlebell from the floor to arm's length during the get-up. These can be done with your knees bent, with straight legs, and also by using the bottom-up position. An additional variation is a bridge floor press where your hips are off the floor and the only contact points are your shoulders and your feet. An additional advantage of performing this exercise is that you are combining two great strength exercises into one—a dynamic and a static exercise together. Other than that, the movement is performed as the floor press.

SINGLE KETTLEBELL CLEAN

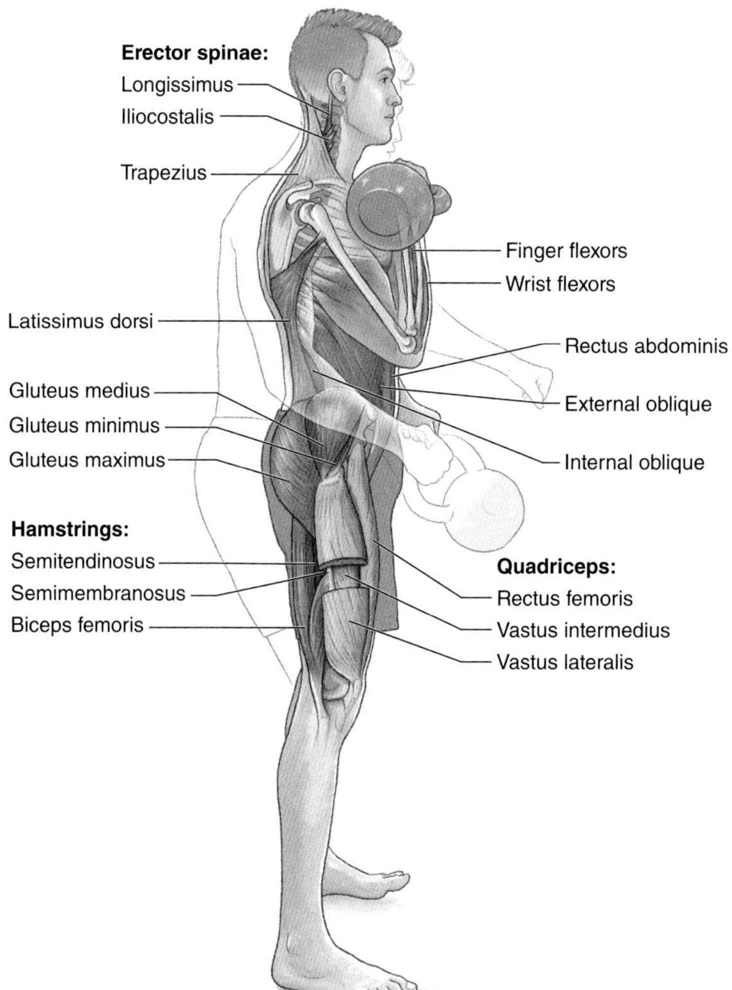

Execution

1. Place your feet about a foot (0.3 m) behind the kettlebell, feet shoulder-width apart with your toes pointed slightly outward.
2. Keeping your hips above your knees and below your shoulders, grip the kettlebell handle with one hand, placing your body into the hip-hinge position. You should feel tension in your hamstrings. Tilt the handle toward you, allowing it to become an extension of your arm.

3. Place your nonworking hand beside the kettlebell but not on it. Doing this will help keep your shoulders square at the beginning.
4. Grip the ground with your toes, tighten the latissimi dorsi, abdominal muscles, and grip, and hike pass the kettlebell back between your legs without changing your hip-hinge position. Target the kettlebell to the small triangle above your knees.
5. As the kettlebell is moving forward after the hike pass, keep your upper arm against your ribs while simultaneously squeezing your gluteal muscles hard, pulling your kneecaps up strongly, and bracing your abdominal muscles hard while standing up. At the same time, bring your elbow back. Perform a kettlebell clean to your waist.
6. Jab your hand through the handle to seat it properly in the rack, or final, position.
7. At the top of the clean, your body will be in a vertical plank position, very similar to the kettlebell swing. Instead of the kettlebell being at arm's length in front of you, you will be able to touch your chest or clavicle with your thumb in the rack position, if needed.
8. Hold this top position briefly. At the top, your shoulders and hip should be squared up.
9. To start the descent of the kettlebell for the next repetition, keep the upper arm against your torso while lowering the kettlebell. Keep your vertical plank position until the last moment, and then get out of the way of the kettlebell. Doing this will keep the kettlebell in the small triangle between your legs during your hip hinge.
10. Complete all repetitions on one side, then switch hands. When you are done, return the kettlebell to the ground, similar to what you would do in the kettlebell swing.

Muscles Involved

Primary: Erector spinae (iliocostalis, longissimus, spinalis), gluteus maximus, gluteus medius, gluteus minimus, quadratus lumborum, hamstrings (semitendinosus, semimembranosus, biceps femoris), latissimus dorsi, brachioradialis, rectus abdominis, external oblique, internal oblique, transversus abdominis

Secondary: Trapezius, rhomboids, quadriceps (rectus femoris, vastus lateralis, vastus medialis, vastus intermedius), biceps brachii, forearms (wrist flexors, finger flexors)

(continued)

SINGLE KETTLEBELL CLEAN *(continued)*

Anatomic Focus

Hand spacing: Contact the middle of the kettlebell handle with one hand, clasping it loosely with your fingers.

Grip: Use a single overhand grip (pronate the hand).

Stance: Position your legs approximately one foot (0.3 m) behind the kettlebell and shoulder-width apart. Point your toes slightly outward.

Trajectory: Bending your elbow and slightly extending the upper arm on the upswing, rotate the body of the kettlebell around the wrist and forearm. The clean is a vertical projection of force, whereas the swing is a horizontal projection of force.

Range of motion: Keeping your arm straight, hike the kettlebell back between your legs, aiming for the small triangle above your knees. Keep your hip-hinge position the same while hiking the kettlebell. When you have reached the end of the hike, stand up quickly while keeping your upper arm against your torso and slightly extending your elbow to be able to jab your hand through the handle and finish in the rack position. Your toes will be rooted to the ground, your kneecaps pulled up, gluteal muscles squeezed, and your abdominal muscles and latissimi dorsi strongly engaged. As in the kettlebell deadlift, the erector spinae, abdominal muscles, and latissimi dorsi will help to stabilize and straighten the spine while the gluteus maximus and hamstrings generate hip extension. Your spine should be straight and stiff throughout the movement. Antishrug your shoulders, further activating your latissimi dorsi. Do not overextend your spine.

During one-hand swings, my recommendation is to have your nonworking arm mimic the movements of the working arm. If you wish, when the kettlebell is at the top of the movement, place the nonworking arm in a guard position instead of having it out in front. Moving the nonworking arm in this way will help avoid rotation of the torso during the exercise.

Technique note: Switch your start hand at the beginning of each succeeding set. Doing this prevents you from having a favorite start hand and will help in keeping a balanced body while training. Women, do not hit your breasts with the kettlebell or your arms.

VARIATION

Dead Stop Clean

This variation is performed the same as the single kettlebell clean except that every repetition starts and stops with the kettlebell on the ground between your feet, similar to the kettlebell deadlift. The top part of this variation is the rack position. This variation teaches you to keep your upper arm against your torso. This exercise is very powerful and explosive and is generally done for three to five repetitions per hand per set.

DOUBLE KETTLEBELL CLEAN

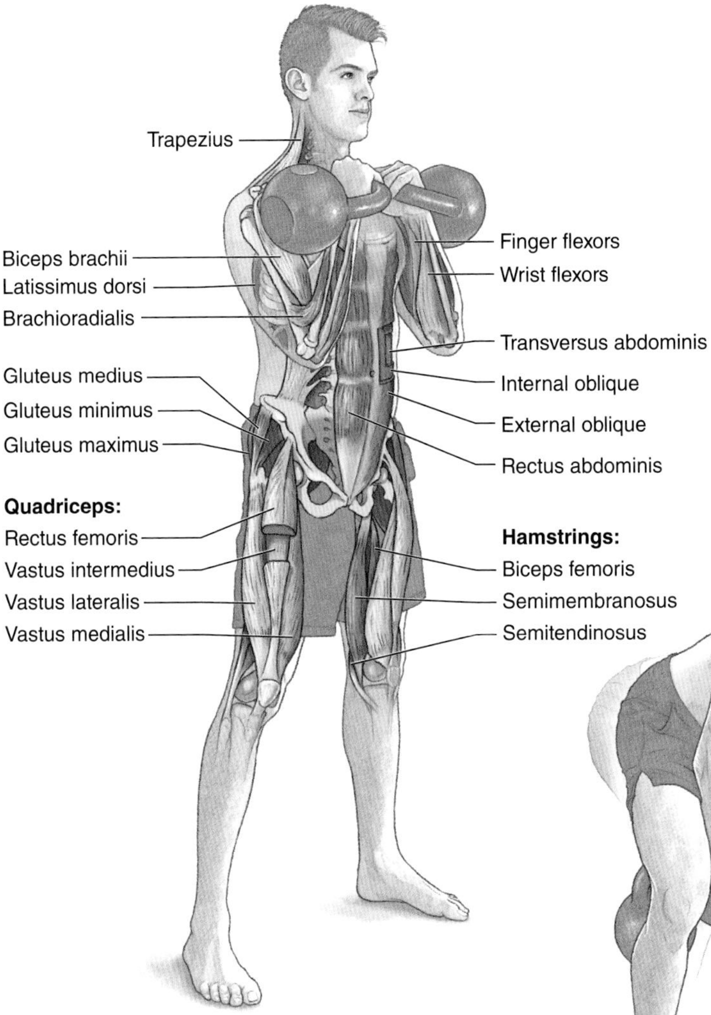

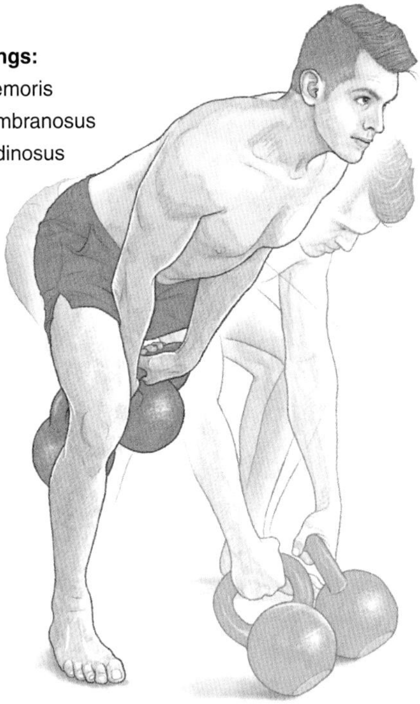

Start position and swing.

Execution

1. Place your feet about a foot (0.3 m) behind the kettlebell, feet slightly more than shoulder-width apart, with your toes angled slightly outward.
2. Keeping your hips above your knees and below your shoulders, grip a kettlebell handle in each hand, placing your body into the hip-hinge position. You should feel tension in your hamstrings. Tilt the handles toward you, allowing them to become an extension of your arms.

(continued)

DOUBLE KETTLEBELL CLEAN *(continued)*

3. Grip the ground with your toes, tighten the latissimi dorsi, abdominal muscles, and grip, and hike pass the kettlebells back between your legs without changing your hip-hinge position. Target the kettlebells to the small triangle above your knees.
4. As the kettlebells are moving forward after the hike pass, keep your upper arms against your ribs while simultaneously squeezing your gluteal muscles hard, pulling your kneecaps up strongly, bracing your abdominal muscles hard, and standing up. At the same time, bring your elbows back. Perform a clean to your waist.
5. Jab your hands through the handles to seat the kettlebells properly in the rack, or final, position.
6. At the top of the clean, your body will be in a vertical plank position, very similar to the kettlebell swing. Instead of the kettlebells being at arm's length in front of you, you will be able to touch your chest or clavicle with your thumbs in the rack position, if needed.
7. Hold this top position briefly. At the top, your shoulders and hip should be squared up.
8. To start the descent of the kettlebells for the next repetition, keep the upper arms against your torso while lowering the kettlebells. Keep your vertical plank position until the last moment, and then get out of the way of the kettlebells. Doing this will keep the kettlebells in the small triangle between your legs during your hip hinge.
9. Complete all repetitions and then set the kettlebells on the ground, as you would do in the kettlebell swing.

Muscles Involved

Primary: Erector spinae (iliocostalis, longissimus, spinalis), gluteus maximus, hamstrings (semitendinosus, semimembranosus, biceps femoris), latissimus dorsi, brachioradialis, rectus abdominis, external oblique, internal oblique, transversus abdominis

Secondary: Trapezius, rhomboids, quadriceps (rectus femoris, vastus lateralis, vastus medialis, vastus intermedius), quadratus lumborum, gluteus medius, gluteus minimus, biceps brachii, forearms (wrist flexors, finger flexors)

Anatomic Focus

Hand spacing: Put each hand in contact with the middle of the kettlebell handle, clasping it loosely with your fingers.

Grip: Use a double overhand grip, with a hand on each kettlebell. Align the kettlebell handles with each other or make a *V* with the handles.

Stance: Position your legs approximately one foot (0.3 m) behind the kettlebell and slightly more than shoulder-width apart. Point your toes slightly outward.

Trajectory: Bending your elbows and slightly extending the upper arms on the upswing, rotate the body of the kettlebells around the wrists and forearms. The clean is a vertical projection of force, whereas the swing is a horizontal projection of force.

Range of motion: Keeping your arms straight, hike the kettlebells back between your legs, aiming for the small triangle above your knees. Keep your hip-hinge position the same while hiking the kettlebells. When you have reached the end of the hike, stand up quickly while keeping your upper arms against your torso while slightly extending your elbows to be able to jab your hands through the handles and finish in the rack position. Your toes will be rooted to the ground, your kneecaps pulled up, your gluteal muscles squeezed, and your abdominal muscles and latissimi dorsi strongly engaged. As in the kettlebell deadlift, the erector spinae, abdominal muscles, and latissimus dorsi muscles will help to stabilize and straighten the spine while the gluteus maximus and hamstrings generate hip extension. Your spine should be straight and stiff throughout the movement. Antishrug your shoulders, further activating your latissimus dorsi muscles. Do not overextend your spine.

Technique note: When you are in the rack position, do not bang your fingers together. Either have your fingers extended or keep the handles separated at the top. Women, do not hit your breasts with the kettlebells or your arms.

SINGLE KETTLEBELL PRESS

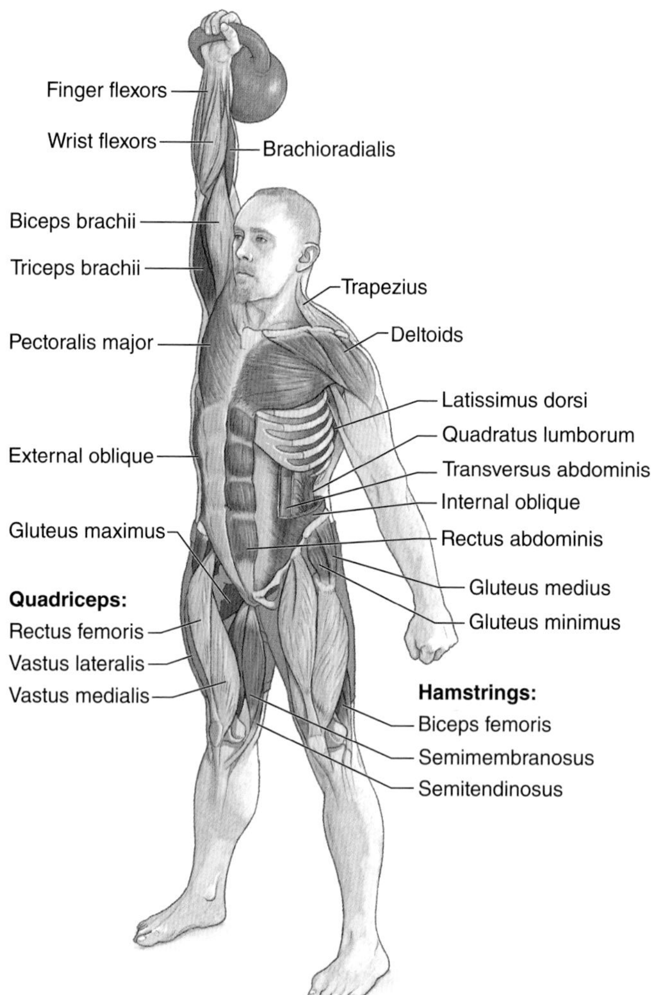

Execution

1. Place your feet about a foot (0.3 m) behind the kettlebell, feet shoulder-width apart with your toes pointed slightly outward. Perform the single kettlebell clean (page 64) and pause in the rack position.
2. Move your forearm to a vertical or near-vertical position.
3. Press up, keeping your humerus at about a 45-degree angle from the front until your hand passes your head. Start to flare your elbow out for the rest of the press.
4. At the top, the kettlebell will be above your head and to the side with your upper arm beside your ear.

5. On the descent, retrace the path down to the rack position. Pause and press again for the required repetitions.
6. Complete all repetitions on one side, then switch hands. When you are done, return the kettlebell to the ground as you would in the kettlebell swing.

Muscles Involved

Primary: Erector spinae (iliocostalis, longissimus, spinalis), gluteus maximus, gluteus medius, gluteus minimus, quadratus lumborum, hamstrings (semitendinosus, semimembranosus, biceps femoris), latissimus dorsi, pectoralis major, deltoids, triceps brachii, brachioradialis, rectus abdominis, external oblique, internal oblique, transversus abdominis

Secondary: Trapezius, rhomboids, quadriceps (rectus femoris, vastus lateralis, vastus medialis, vastus intermedius), biceps brachii, forearms (wrist flexors, finger flexors)

Anatomic Focus

Hand spacing: Contact the middle of the kettlebell handle with one hand, clasping it loosely with your fingers.

Grip: Use a single overhand grip (pronate the hand).

Stance: Position your legs approximately one foot (0.3 m) behind the kettlebell and shoulder-width apart. Point your toes slightly outward.

Trajectory: After performing a kettlebell clean, it will be traveling a slight diagonal path up to the lockout position with the elbow leading the way.

Range of motion: A strong clean means the foundation for a strong press. The press is a grind—a strength movement. You are going from a dead stop to overhead and back again. The first repetition will be a concentric-only move, so make sure that you have your toes rooted to the ground, your kneecaps pulled up, your gluteal muscles squeezed, and your abdominal muscles and latissimi dorsi all strongly engaged. Use only enough tension to get the job done. As in the kettlebell deadlift, the erector spinae, abdominal muscles, gluteus maximus, and latissimus dorsi muscles will help to stabilize the spine. Your spine should be straight and stiff throughout the movement. Antishrug your shoulders, further activating your latissimus dorsi muscles. Do not overextend your spine, especially at the top.

With your nonworking hand during the single press, make a strong fist at the same time as you are crush gripping the handle of the kettlebell at the start of the press. This further tenses your whole body and adds to the strength available to lock out the press. Relax the hands in the rack position.

Technique note: Press with your elbow, not with your hand. In other words, on the way up think about moving your elbow from the flexed to the extended position. The hand will follow.

DOUBLE KETTLEBELL PRESS

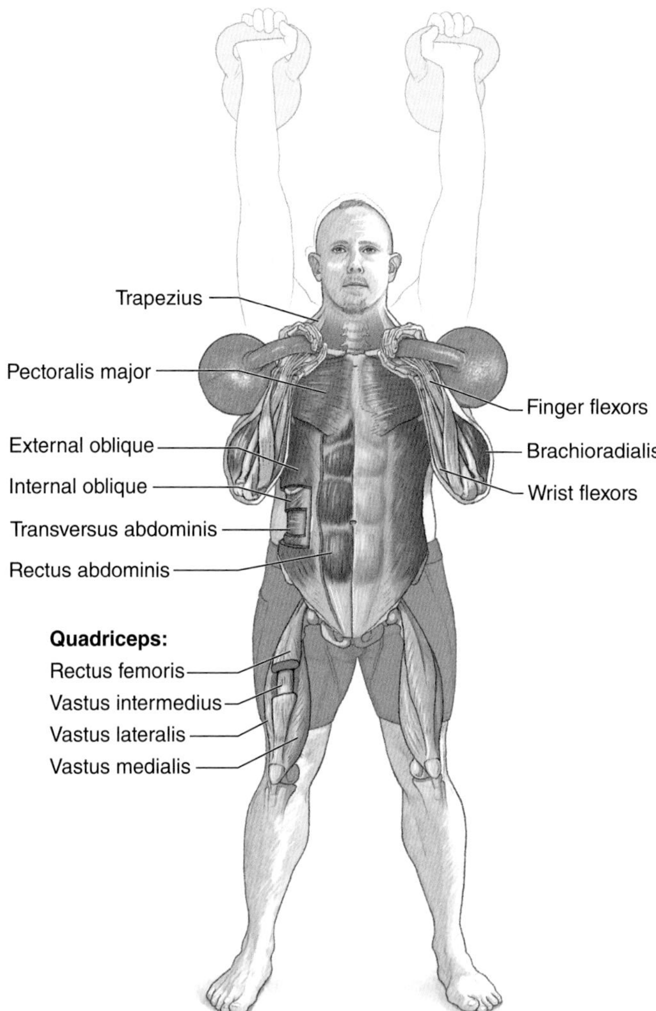

Execution

1. Place your feet about a foot (0.3 m) behind the kettlebells, shoulder-width apart with your toes pointed slightly outward. Perform the double kettlebell clean (page 67) and pause in the rack position.
2. Move your forearms to a vertical or near-vertical position.
3. Press up, keeping both humerus bones at about a 45-degree angle from the front until your hands pass your head. Start to flare your elbows out for the rest of the press.
4. At the top, the kettlebells will be above your head and to the side with your upper arms beside your ears.

5. On the descent, retrace the path down to the rack position. Pause and press again for the required repetitions.
6. Complete all repetitions and then set the kettlebells on the ground, as you would do in the kettlebell swing.

Muscles Involved

Primary: Erector spinae (iliocostalis, longissimus, spinalis), gluteus maximus, hamstrings (semitendinosus, semimembranosus, biceps femoris), latissimus dorsi, pectoralis major, deltoids, triceps brachii, brachioradialis, rectus abdominis, external oblique, internal oblique, transversus abdominis

Secondary: Trapezius, rhomboids, quadriceps (rectus femoris, vastus lateralis, vastus medialis, vastus intermedius), gluteus medius, gluteus minimus, quadratus lumborum, biceps brachii, forearms (wrist flexors, finger flexors)

Anatomic Focus

Hand spacing: Contact the middle of the kettlebell handle with each hand, clasping it loosely with your fingers.

Grip: Use a double overhand grip with a hand on each kettlebell. Align the kettlebell handles with each other or make a *V* with the handles.

Stance: Position your feet approximately one foot (0.3 m) behind the kettlebell and shoulder-width apart. Point your toes slightly outward.

Trajectory: After performing kettlebell cleans, the kettlebells will be traveling a slight diagonal path up to the lockout position with the elbows leading the way.

Range of motion: A strong clean is the foundation for a strong press. The press is a grind—a strength movement. You are going from a dead stop to overhead and back again. The first repetition will be a concentric-only move, so make sure that you have your toes rooted to the ground, your kneecaps pulled up, your gluteal muscles squeezed, and your abdominal muscles and latissimi dorsi all strongly engaged. Crush grip the handles of the kettlebells at the start of the lift. Relax the hands in the rack position. Use only enough tension to get the job done. As in the kettlebell deadlift, the erector spinae, abdominal muscles, gluteus maximus, and latissimus dorsi muscles will help to stabilize the spine. Your spine should be straight and stiff throughout the movement. Antishrug your shoulders from the start of the press until the end, further activating your latissimus dorsi muscles. Do not overextend your spine, especially at the top.

Technique note: Press with your elbows, not with your hands. On the way up think about moving your elbows from the flexed to the extended position. The hands will follow.

When you are in the rack position, do not bang your fingers together. Either have your fingers extended or keep the handles separated at the top. Women, do not hit your breasts with the kettlebell or your arms.

(continued)

DOUBLE KETTLEBELL PRESS *(continued)*

VARIATIONS

Alternating Press

This variation is performed the same as the double kettlebell press except instead of pressing both kettlebells at the same time, you press one kettlebell to the top and back to the rack position before you press the other kettlebell. Keep the vertical plank position throughout the set.

See-Saw Press

This variation is performed the same as the double kettlebell press with one change: press one kettlebell all the way up. On the descent of that kettlebell, start ascending with the other kettlebell, your hands meeting approximately at the top of your head. A slight side bend is allowed for this press while keeping the vertical plank position throughout the set.

SOTS PRESS WITH DOUBLE KETTLEBELLS

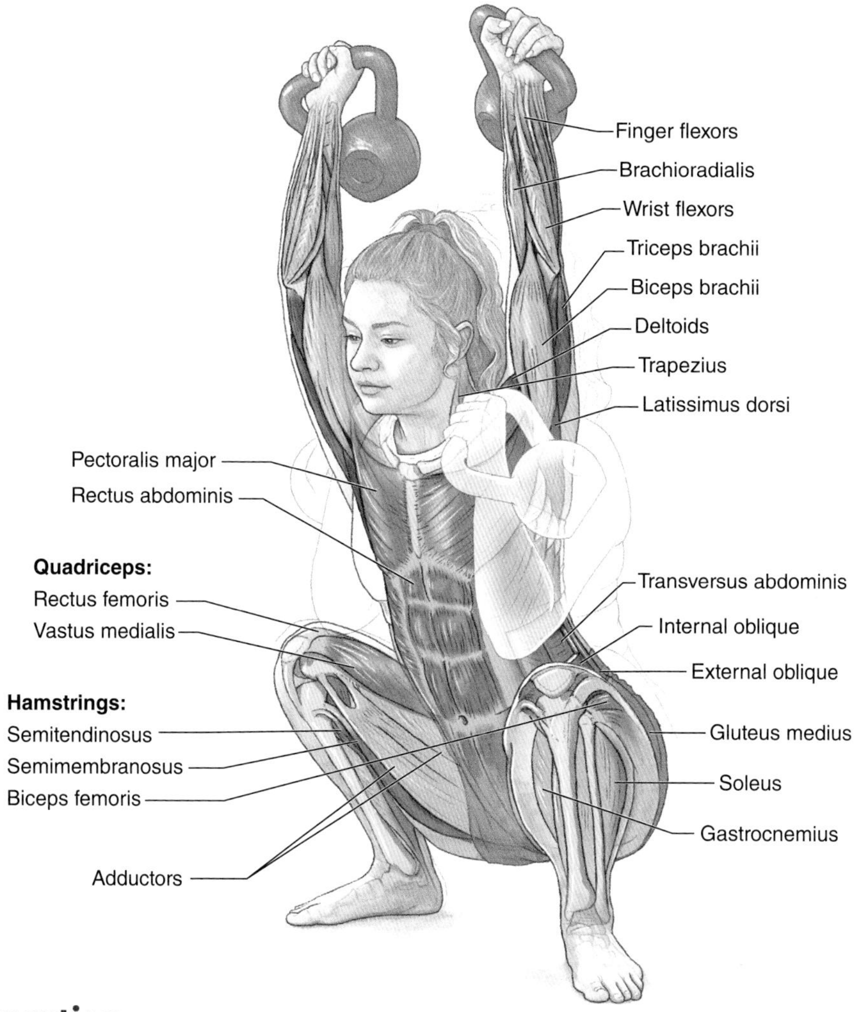

Finger flexors
Brachioradialis
Wrist flexors
Triceps brachii
Biceps brachii
Deltoids
Trapezius
Latissimus dorsi
Pectoralis major
Rectus abdominis
Quadriceps:
Rectus femoris
Vastus medialis
Transversus abdominis
Internal oblique
External oblique
Hamstrings:
Semitendinosus
Semimembranosus
Biceps femoris
Gluteus medius
Soleus
Gastrocnemius
Adductors

Execution

1. Perform the double kettlebell clean (page 67). Assume your squat stance: toes pointed out, knees in line with the middle of each foot.
2. Lower your hips to squat down to where you are comfortable (preferably below parallel, with your hips below your knees) and pause.
3. Press both kettlebells up to the locked position overhead. Make sure your upper extremities are perpendicular to the ground.
4. Slowly and under control, return the kettlebells to the rack position and repeat.
5. When done with the repetitions, stand up, assume a double-clean stance, and return the kettlebells to the ground.

(continued)

SOTS PRESS WITH DOUBLE KETTLEBELLS *(continued)*

Muscles Involved

Primary: Erector spinae (iliocostalis, longissimus, spinalis), quadriceps (rectus femoris, vastus lateralis, vastus medialis, vastus intermedius), gluteus maximus, hamstrings (semitendinosus, semimembranosus, biceps femoris), latissimus dorsi, pectoralis major, deltoids, triceps brachii, rectus abdominis, external oblique, internal oblique, transversus abdominis

Secondary: Trapezius, rhomboids, gluteus medius, gluteus minimus, abductors and adductors, calf muscles (gastrocnemius, soleus), forearms (wrist flexors, finger flexors), elbow flexors (biceps brachii, brachioradialis)

Anatomic Focus

Hand spacing: Contact the middle of the kettlebell handle with each hand, clasping it loosely with your fingers.

Grip: Use a double overhand grip. A hand will be on each kettlebell. Align the kettlebell handles with each other or make a V with the handles.

Stance: Position the legs approximately shoulder-width apart. Point your toes slightly outward.

Trajectory: Move the kettlebells in a slight diagonal path up to the lockout position with the elbow leading the way during the press portion.

Range of motion: A strong clean means the foundation for a strong squat followed by a strong press. The press is a grind—a strength movement. You are going from a dead stop to overhead and back again while in a paused squat motion. The first repetition will be a concentric-only move, so make sure that you have your toes rooted to the ground and your abdominal muscles and latissimi dorsi are strongly engaged. Crush grip the handles of the kettlebells at the start of the lift. Relax the hands in the rack position. This is an advanced kettlebell movement. It should be done by those that have adequate mobility to perform the exercise. The increased core stabilization will further help you in your athletic endeavors or handling the rigors of everyday life. Your spine should be straight and stiff throughout the movement. Antishrug your shoulders from the start of the press until the end, further activating your latissimus dorsi muscles. The Sots press can be performed as a warm-up, as a main training exercise, or as a specialized variation.

Technique note: Press with your elbows, not with your hands. What this means is that on the way up, think about moving your elbows from the flexed to the extended position. The hands will follow.

VARIATIONS

Alternating Sots Press

This variation is performed the same way as the Sots press with double kettlebells except instead of pressing both kettlebells at the same time, you press one kettlebell to the top and back to the rack position before you start the other kettlebell. Maintain the neutral spine position and keep your midsection tight.

See-Saw Sots Press

This variation is performed the same way as the Sots press with double kettlebells with one change: Press one kettlebell all the way up. On the descent of that kettlebell, start ascending with the other kettlebell. Your hands will meet approximately at the top of your head. A slight side bend is allowed for this press while maintaining the neutral spine position and keeping your midsection tight.

SINGLE KETTLEBELL FLOOR PRESS

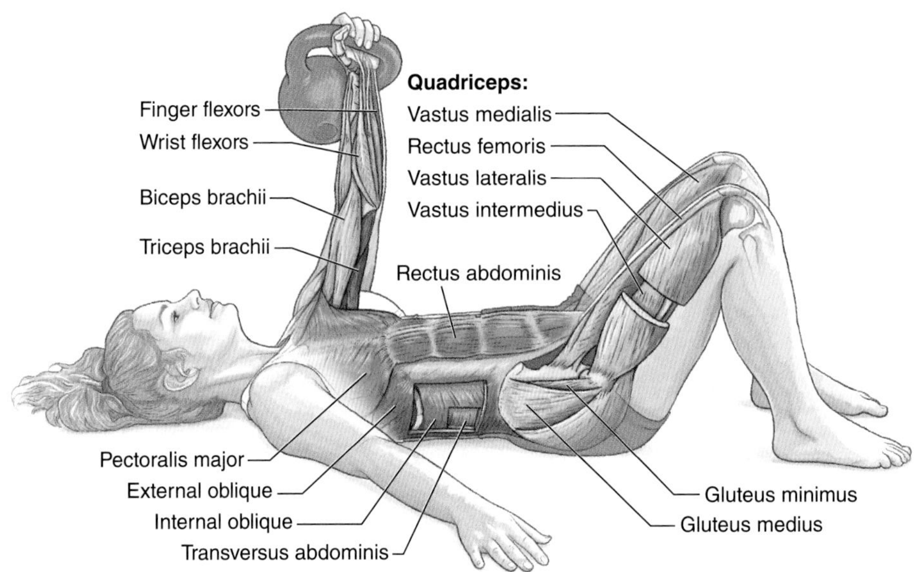

Execution

1. Lie on your left side with the kettlebell close to you. Grab the handle with your left hand and place your right hand with a thumbless grip over the top. While keeping the kettlebell close to you, roll to your back with the two-handed grip. The kettlebell will be down by your ribs with your left elbow on the ground. Let go with your right hand and place it on the floor.
2. Keep your knees bent and your feet flat on the floor.
3. Grip the ground with your toes, squeeze your gluteal muscles, flatten your back, tighten your abdominal muscles, and crush grip the kettlebell handle. Straighten the arm as you press the kettlebell toward the ceiling.
4. Lower the kettlebell by bending your arm at the elbow until the back side of your humerus touches the floor, pause briefly, and press it back up.
5. Keep your humerus between a 0- and 45-degree-angle from your ribs. Your hand will also be at the same angle.

6. Complete the repetitions. Reverse the movement that you performed to put the kettlebell into position for the floor press.
7. When switching sides, you can use both hands to drag the kettlebell on the floor around your head in an arc to the other side or move your body to the other side of the kettlebell.

Muscles Involved

Primary: Pectoralis major, latissimus dorsi, deltoids, triceps brachii, rectus abdominis, external oblique, internal oblique, transversus abdominis

Secondary: Erector spinae (iliocostalis, longissimus, spinalis), gluteus maximus, trapezius, rhomboids, quadriceps (rectus femoris, vastus lateralis, vastus medialis, vastus intermedius), gluteus medius, gluteus minimus, biceps brachii, forearms (wrist flexors, finger flexors)

Anatomic Focus

Grip: Hold the kettlebell handle in a diagonal position from the web of the thumb to the pisiform of the hand. The wrist will be neutral.

Stance: Assume a supine position with your knees bent and your feet flat on the floor.

Trajectory: Move the kettlebell in a vertical motion. A slight midline motion of the kettlebell on the ascent is acceptable if desired.

Range of motion: Essentially, you have taken part of the get-up exercise and created a new one: a floor press. During this movement you are still using your whole body with less overall movement. This is a grind exercise, therefore pay attention to the motion of the kettlebell up and down. When done properly, it will help make your shoulders and upper body bulletproof and make your torso strong, especially in activities that require a pressing action. Prior to moving the kettlebell, check that you are in the best position to succeed before performing it.

Technique note: This exercise not only helps develop both sides of the body but also increases the upper arm strength. With the limited range of motion of this movement, this is a great exercise for those athletes with shoulder issues.

This movement can be a main exercise, or it can be used as specialized variety supplementing your upper body and torso strength.

(continued)

SINGLE KETTLEBELL FLOOR PRESS *(continued)*

VARIATIONS

Straight Legs

The main difference between this variation and the single kettlebell floor press is that your legs will be straight. Prior to pressing the kettlebell up, flatten your back and brace your abdominal muscles. Squeeze your gluteal muscles and keep both heels on the ground. This variation works your torso more than the former exercise and is harder overall. Spend some time with the single kettlebell floor press before attempting this variation.

Bottom-Up Press

The main difference between this variation and the single kettlebell floor press is that you will place the kettlebell into a bottom-up position prior to pressing it up. Move the kettlebell slowly through the range of motion, both on the ascent and the descent. Prior to pressing the kettlebell up, crush grip the handle, flatten your back, and brace your abdominal muscles. Squeeze your gluteal muscles and keep both feet on the ground. Spend some time with the single kettlebell floor press before attempting this variation.

DOUBLE KETTLEBELL FLOOR PRESS

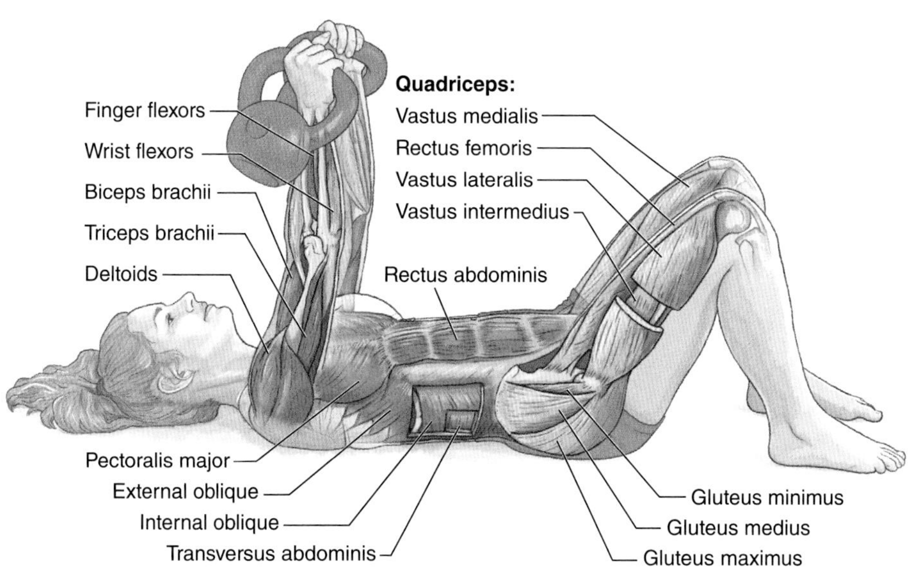

Execution

1. Place two kettlebells on the ground, one on each side of your shoulders. Lie down with your legs in front of you. Roll to your side in the fetal position and pick up one of the kettlebells as you would for the single kettlebell floor press (page 78).
2. While keeping your weighted hand and forearm against your chest, roll to the other side and pick up the other kettlebell with your other hand and roll to your back. Make sure your abdominal muscles are tight and braced throughout this part of the execution. Your humerus bones will be against the sides of your ribs with your hands in a neutral position.
3. Move your humerus bones to the desired angles from your torso—angled 0 to 45 degrees to your ribs.
4. Make sure your knees are bent and your feet are flat on the floor.
5. Grip the ground with your toes, squeeze your gluteal muscles, flatten your back, tighten your abdominal muscles, and crush grip the kettlebell handles. Straighten your arms as you press the kettlebells toward the ceiling.
6. Lower the kettlebells by bending your arms at the elbows until the back side of your humerus bones touch the floor; pause briefly and press the kettlebells back up.

(continued)

DOUBLE KETTLEBELL FLOOR PRESS *(continued)*

7. Keep your humerus bones angled 0 to 45 degrees from your ribs during the ascent and descent. Your hands will be at the same angle.
8. Complete the repetitions. Reverse the movement that you performed to put the kettlebells into position for the floor press. Make sure your abdominal muscles are tight and braced while placing the kettlebells on the floor.

Muscles Involved

Primary: Pectoralis major, latissimus dorsi, deltoids, triceps brachii, biceps brachii, rectus abdominis, external oblique, internal oblique, transversus abdominis

Secondary: Erector spinae (iliocostalis, longissimus, spinalis), gluteus maximus, trapezius, rhomboids, quadriceps (rectus femoris, vastus lateralis, vastus medialis, vastus intermedius), gluteus medius, gluteus minimus, forearms (wrist flexors, finger flexors)

Anatomic Focus

Grip: Hold the kettlebell handles in a diagonal position from the web of the thumb to the pisiform of the hand.

Stance: Assume a supine position with your knees bent and your feet flat on the floor.

Trajectory: Move the kettlebells in a vertical motion. You may also have a slight midline motion of the kettlebells on the ascent.

Range of motion: Essentially you have taken part of the get-up exercise and created a new one: a floor press. During this movement, you are still using your whole body with less overall movement. This is a grind exercise; therefore pay attention to the motion of the kettlebells up and down. When done properly, it will help make your shoulders and upper body bulletproof and make your torso strong, especially in activities that require a pressing action. Prior to moving the kettlebells, check that you are in the best position to succeed before performing it.

Technique note: This exercise not only helps develop both sides of the body but also increases upper-arm strength. With the limited range of motion of this movement, this is a great exercise for those athletes with shoulder issues. Practice this exercise with lighter kettlebells before attempting heavier kettlebells.

This movement can either be a main exercise or used as specialized variety supplementing your upper body and torso strength.

VARIATIONS

Straight Legs

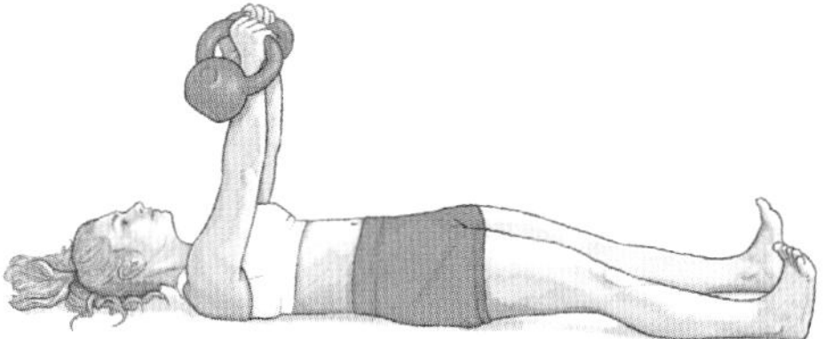

The main difference between this variation and the double kettlebell floor press is that your legs will be straight. Prior to pressing the kettlebells up, flatten your back and brace your abdominal muscles. Squeeze your gluteal muscles and keep both feet on the ground. This variation works your torso more than the double kettlebell floor press and is harder overall. Spend some time with the double kettlebell floor press movement before attempting this variation.

Alternating Press

In this variation of the double kettlebell floor press, press one kettlebell up and pull it back down to the ground before pressing the other kettlebell. Do this for the prescribed number of repetitions. Spend some time with the double kettlebell floor press movement before attempting this variation.

DOUBLE KETTLEBELL BRIDGE FLOOR PRESS

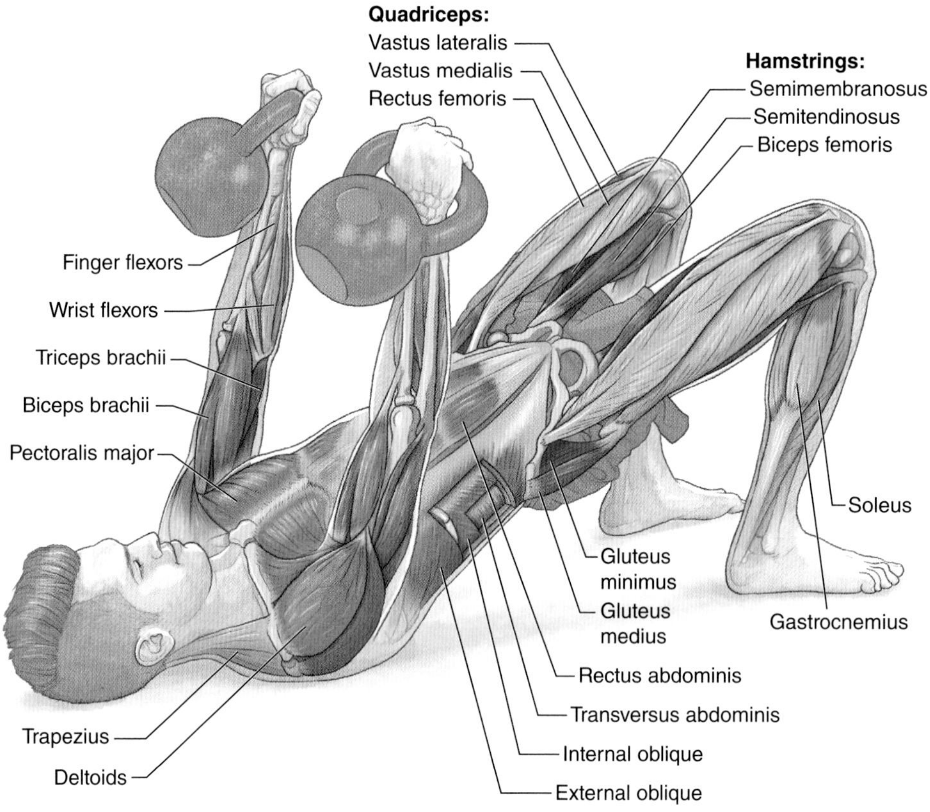

Quadriceps:
Vastus lateralis
Vastus medialis
Rectus femoris

Hamstrings:
Semimembranosus
Semitendinosus
Biceps femoris

Finger flexors
Wrist flexors
Triceps brachii
Biceps brachii
Pectoralis major
Trapezius
Deltoids

Soleus
Gluteus minimus
Gluteus medius
Gastrocnemius
Rectus abdominis
Transversus abdominis
Internal oblique
External oblique

Start position and bridge.

Execution

1. Place two kettlebells on the ground, one on each side of your shoulders. Lie down with your legs in front of you. Roll to your side in the fetal position and pick up one of the kettlebells as you would for the single kettlebell floor press (page 78).
2. While keeping your weighted hand and forearm against your chest, roll to the other side and pick up the other kettlebell with your other hand and roll to your back. Make sure your abdominal muscles are tight and braced throughout this part of the execution. Your humerus bones will be against the sides of your ribs with your hands in a neutral position.
3. Move your humerus bones to the desired angles from your torso—angled 0 to 45 degrees to your ribs.
4. Make sure your knees are bent and your feet are flat on the floor.
5. Flatten your back, tighten your abdominal muscles, and lift your hips off the ground. Grip the ground with your toes, squeeze your gluteal muscles, and crush grip the kettlebell handles.
6. Straighten your arms as you press the kettlebells toward the ceiling.
7. Lower the kettlebells by bending your arms at the elbows until the back sides of your humerus bones touch the floor, pause briefly, and press the kettlebells back up.
8. Keep your humerus bones and hands angled 0 to 45 degrees from your ribs. Your hands will be at the same angle.
9. Complete the repetitions. Lower your hips to the ground and reverse the movement that you performed to put the kettlebells into position for the floor press. Make sure your abdominal muscles are tight and braced while placing the kettlebells on the floor.

Muscles Involved

Primary: Pectoralis major, latissimus dorsi, deltoids, triceps brachii, biceps brachii, gluteus maximus, gluteus medius, gluteus minimus, hamstrings (semitendinosus, semimembranosus, biceps femoris), rectus abdominis, external oblique, internal oblique, transversus abdominis

Secondary: Erector spinae (iliocostalis, longissimus, spinalis), trapezius, rhomboids, quadriceps (rectus femoris, vastus lateralis, vastus medialis, vastus intermedius), gastrocnemius, soleus, forearms (wrist flexors, finger flexors)

(continued)

DOUBLE KETTLEBELL BRIDGE FLOOR PRESS *(continued)*

Anatomic Focus

Grip: Hold the kettlebell handles in a diagonal position from the web of the thumb to the pisiform of the hand.

Stance: Assume a supine position with your hips raised, your knees bent, and your feet flat on the floor.

Trajectory: Move the kettlebells in a vertical motion. Effect a slight midline motion of the kettlebells on the ascent if desired.

Range of motion: This movement combines two great moves into one: the floor bridge and the kettlebell floor press. The floor press is dynamic, and the floor bridge is static. The exercise is essentially done the same as the double kettlebell floor press except your hips will be raised off the floor during the repetitions of the floor press. Make certain that your lower back is flat, and your abdominal muscles are braced before lifting your hips off the floor. This helps to prevent extending the lumbar spine during the exercise.

During this movement you are using your whole body with less overall movement. This is a grind exercise; therefore, pay attention to the motion of the kettlebells up and down. When done properly, it will help make your shoulders and upper body bulletproof and make your torso and hips strong. Prior to moving the kettlebells, check that you are in the best position to succeed before performing the exercise. With the limited range of motion of this movement, this is great exercise for those athletes with shoulder issues. This movement can either be a main exercise or used as specialized variety supplementing your upper and lower body and torso strength.

VARIATION

Alternating Press

In this variation of the double kettlebell bridge floor press, you will press up one kettlebell and pull it back down to the ground before pressing the other kettlebell. Do this for the prescribed number of repetitions. Spend some time with the double kettlebell bridge floor press movement before attempting this variation.

SINGLE KETTLEBELL BOTTOM-UP CLEAN

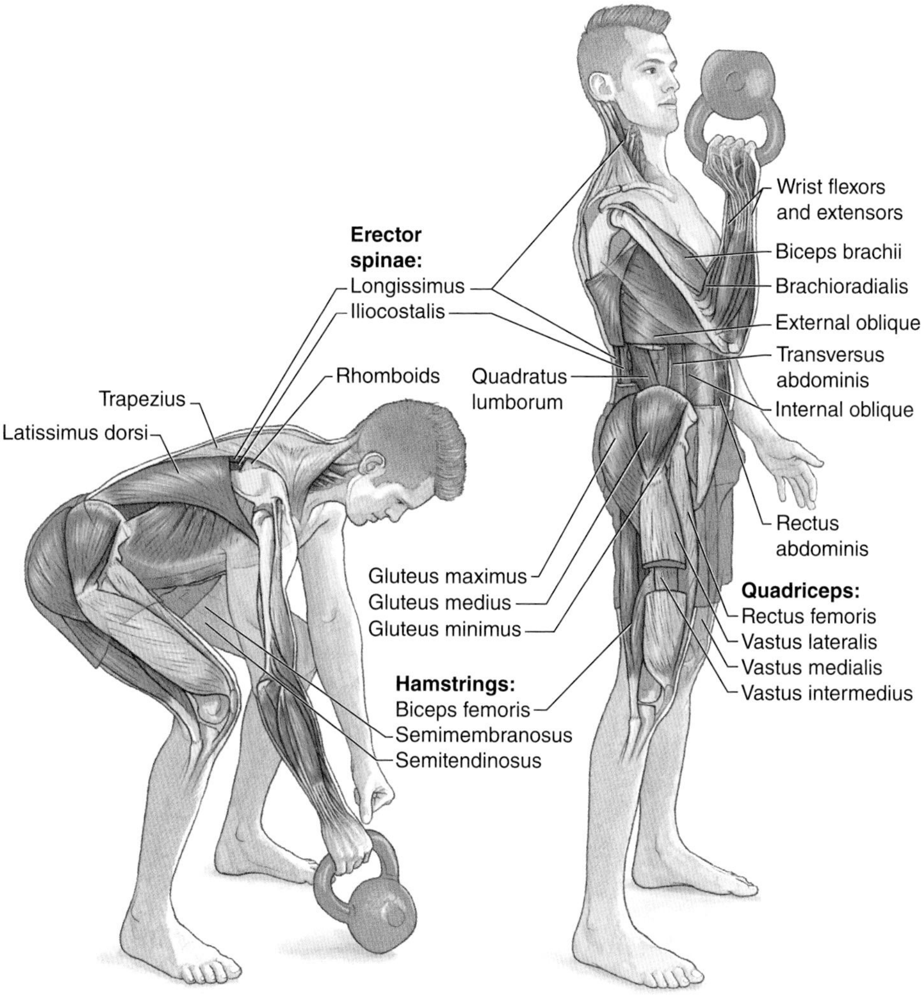

Execution

1. Place your feet about a foot (0.3 m) behind the kettlebell, with your feet shoulder-width apart with your toes pointed slightly outward. Turn the kettlebell handle perpendicular to your shoulders.
2. Keeping your hips above your knees and below your shoulders, grip the kettlebell handle with one hand, placing your body into the hip-hinge position. Crush gripping the kettlebell is necessary at this point. You should also feel tension in your hamstrings. Tilt the handle toward you, allowing it to become an extension of your arm.

(continued)

SINGLE KETTLEBELL BOTTOM-UP CLEAN *(continued)*

3. Place your nonworking hand beside the kettlebell but not on it. Doing this will help keep your shoulders square at the beginning.
4. Grip the ground with your toes, tighten your latissimi dorsi, abdominal muscles, and grip, and hike pass the kettlebell back between your legs without changing your hip-hinge position. Target the kettlebell to the small triangle above your knees.
5. As the kettlebell is moving forward after the hike pass, keep your upper arm against your ribs while simultaneously squeezing your gluteal muscles hard, pulling your kneecaps up strongly, bracing your abdominal muscles hard, and standing up. At the same time, while flexing your elbow, bring it under the bottom-up kettlebell. Perform a kettlebell clean to your waist.
6. At the top of the clean, your body will be in a vertical plank position, very similar to the kettlebell swing position. Your elbow and wrist will be below the kettlebell in proper alignment. Try to deform the kettlebell handle by crush gripping it.
7. Hold this top position briefly. At the top, square up your shoulders and hip.
8. To start the descent of the kettlebell for the next repetition, keep the upper arm against your torso while lowering the kettlebell. Keep your vertical plank position until the last moment, and then get out of the way of the kettlebell. Doing this will keep the kettlebell in the small triangle between your legs during your hip hinge.
9. Complete all repetitions on one side, then switch hands. When done, return it to the ground similar to what is done in the kettlebell swing.

Muscles Involved

Primary: Erector spinae (iliocostalis, longissimus, spinalis), gluteus maximus, contralateral quadratus lumborum, gluteus medius, gluteus minimus, hamstrings (semitendinosus, semimembranosus, biceps femoris), latissimus dorsi, brachioradialis, rectus abdominis, external oblique, internal oblique, transversus abdominis, forearms (wrist flexors and extensors, finger flexors)

Secondary: Trapezius, rhomboids, quadriceps (rectus femoris, vastus lateralis, vastus medialis, vastus intermedius), ipsilateral gluteus medius and gluteus minimus, biceps brachii

Anatomic Focus

Hand spacing: Contact the middle of the kettlebell handle with one hand, clasping it with a crush grip.

Grip: Use a single overhand grip (pronate the hand).

Stance: Position your legs approximately one foot (0.3 m) behind the kettlebell and shoulder-width apart. Point your toes slightly outward.

Trajectory: The movement of the kettlebell during the bottom-up clean is like the movement of the single clean; however, the body of the kettlebell will be above the hand, as you will be crush gripping the handle. The clean is a vertical projection of force whereas the swing is a horizontal projection of force.

Range of motion: The overall motion of the single kettlebell bottom-up clean is almost identical to the motion in the single kettlebell clean, except that from the floor until the end of the clean at the top, you will be crush gripping the handle of the kettlebell. At the top you will be tensing every muscle in your body, frozen in time. Hold this position for a few seconds and then perform a hike pass, relaxing your muscles for a brief time, and repeat.

At the top, keep your upper arm against your torso and antishrug hard. Your toes will be rooted to the ground, your kneecaps pulled up, gluteal muscles squeezed, and your abdominal muscles and latissimi dorsi strongly engaged. As in the kettlebell deadlift, the erector spinae, abdominal muscles, and latissimus dorsi muscles will help to stabilize and straighten the spine while the gluteus maximus and hamstrings generate hip extension. Your spine should be straight and stiff throughout the movement. Antishrug your shoulders, further activating your latissimus dorsi muscles. Do not overextend your spine.

During this kettlebell clean, my recommendation is to place the nonworking arm in a guard position at the top of the clean. Doing this will help avoid rotation of the torso during the exercise.

Technique note: When learning the bottom-up clean, have patience. Perform a few repetitions, maybe two or three per side. You will be generating a lot of tension with each repetition; therefore, you definitely want quality over quantity here. Your abdominal and gluteal muscles need to be cramped and tight to protect your lower back during bottom-up exercises. If you lose your grip on the kettlebell, make sure to get out of the way of the falling kettlebell. Practice in a space where dropping the kettlebell won't hurt the floor. You will find that one side of your body will be weaker than the other side. Hard work here pays off.

(continued)

SINGLE KETTLEBELL BOTTOM-UP CLEAN *(continued)*

VARIATION

Single Kettlebell Bottom-Up Press

This variation is performed as the regular kettlebell press except the kettlebell is bottom up. Start with the single bottom-up clean and move on to the press. During the press, move the kettlebell slowly upward, keeping your wrist and elbow directly below the upside-down kettlebell. Keep using your latissimi dorsi, both in the eccentric and concentric phases of the press. Crush grip the handle even stronger, keeping the kettlebell bottom-up. You will want to lower the kettlebell slowly as well. Remember to keep the number of repetitions on the low end, perhaps as few as two or three.

With your nonworking hand during the single kettlebell bottom-up press, make a strong fist at the same time as the start of the press. Doing this further tenses your whole body and adds to the strength available to lock out the press. Relax the nonworking hand in the rack position.

DOUBLE KETTLEBELL BOTTOM-UP CLEAN

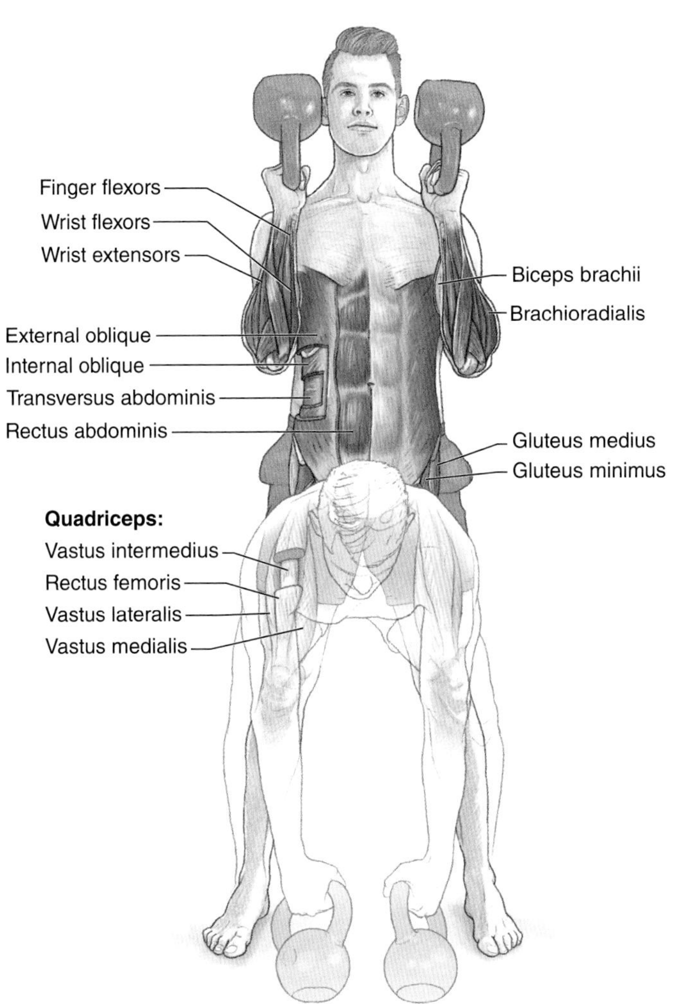

Execution

1. Place your feet about a foot (0.3 m) behind the kettlebells, feet shoulder-width apart with your toes pointed slightly outward. Turn the kettlebell handles perpendicular to your shoulders.
2. Keeping your hips above your knees and below your shoulders, grip the kettlebell handles with each hand, placing your body into the hip-hinge position. Crush gripping the kettlebells is necessary at this point. You should also feel tension in your hamstrings. Tilt the handles toward you, allowing them to become extensions of your arms.

(continued)

DOUBLE KETTLEBELL BOTTOM-UP CLEAN *(continued)*

3. Grip the ground with your toes, tighten the latissimi dorsi, abdominal muscles, and grip, and hike pass the kettlebells back between your legs without changing your hip-hinge position. Target the kettlebells to the small triangle above your knees.
4. As the kettlebells are moving forward after the hike pass, keep your upper arm against your ribs while simultaneously squeezing your gluteal muscles hard, pulling your kneecaps up strongly, bracing your abdominal muscles hard, and standing up. At the same time, while flexing your elbows, bring them under the bottom-up kettlebells. Perform a kettlebell clean to your waist.
5. At the top of the clean, your body will be in a vertical plank position, very similar to the kettlebell swing position. Your elbow and wrist will be below the kettlebells in proper alignment. Imagine trying to deform the kettlebell handles by crush gripping them.
6. Hold this top position briefly. At the top, your shoulders and hip should be squared up.
7. To start the descent of the kettlebells for the next repetition, keep the upper arm against your torso while lowering the kettlebells. Keep your vertical plank position until the last moment, and then get out of the way of the kettlebells. Doing this will keep the kettlebells in the small triangle between your legs during your hip hinge.
8. Complete all repetitions. When done, set the kettlebells on the ground, similar to what is done in the kettlebell swing.

Muscles Involved

Primary: Erector spinae (iliocostalis, longissimus, spinalis), gluteus maximus, hamstrings (semitendinosus, semimembranosus, biceps femoris), latissimus dorsi, brachioradialis, rectus abdominis, external oblique, internal oblique, transversus abdominis, forearms (wrist flexors and extensors, finger flexors)

Secondary: Trapezius, rhomboids, quadriceps (rectus femoris, vastus lateralis, vastus medialis, vastus intermedius), quadratus lumborum, gluteus medius, gluteus minimus, biceps brachii

Anatomic Focus

Hand spacing: Contact the middle of the kettlebell handle with each hand, crush gripping it.

Grip: Use a double overhand grip, with a hand on each kettlebell.

Stance: Position your legs approximately one foot (0.3 m) behind the kettlebells and slightly more than shoulder-width apart. Point your toes slightly outward.

Trajectory: Move the kettlebells during the double kettlebell bottom-up clean as you do in the single kettlebell bottom-up clean. The bodies of the kettlebells will be above each hand, as you will be crush gripping the handles. The clean is a vertical projection of force, whereas the swing is a horizontal projection of force.

Range of motion: The overall motion of the double kettlebell bottom-up clean is almost identical to the motion in the single kettlebell bottom-up clean. From the floor until the end of the clean at the top, you will be crush gripping the handles of the kettlebells. At the top, you will be tensing every muscle in your body. Hold this position for a few seconds and then perform a hike pass, relaxing your muscles for a brief time, and repeat.

At the top, keep your upper arm against your torso and antishrug hard. Root your feet to the ground, pull up your kneecaps, squeeze your gluteal muscles, and strongly engage your abdominal muscles and latissimi dorsi. As in the kettlebell deadlift, the erector spinae, abdominal muscles, and latissimus dorsi muscles will help to stabilize and straighten the spine while the gluteus maximus and hamstrings generate hip extension. Your spine should be straight and stiff throughout the movement. Antishrug your shoulders, further activating your latissimus dorsi muscles. Do not overextend your spine.

Technique note: When learning the double kettlebell bottom-up clean, have patience. Perform a few repetitions, such as two or three per side. You will be generating a lot of tension with each repetition; therefore, you want quality over quantity. Your abdominal and gluteal muscles need to be cramped and tight to protect your lower back during bottom-up exercises. If you lose your grip on the kettlebells, make sure to get out of the way of the falling kettlebells. Practice in a space where dropping the kettlebells won't hurt the floor. You will find that one side of your body will be weaker than the other side. Hard work here to correct this pays off.

(continued)

DOUBLE KETTLEBELL BOTTOM-UP CLEAN *(continued)*

VARIATION

Double Kettlebell Bottom-Up Press

This variation is performed as the regular kettlebell press except the kettlebells are bottom up. Start with the double kettlebell bottom-up clean and move on to the press. During the press, move the kettlebells slowly upward, keeping your wrists and elbows directly below the upside-down kettlebells. Keep using your latissimi dorsi, both in the eccentric and concentric phases of the press. Crush grip the handles even stronger, keeping the kettlebells bottom up. You will want to lower the kettlebells slowly as well. Remember to keep the number of repetitions on the low end, such as two or three. An alternative before pressing both kettlebells up at the same time is to alternate press sides, either redoing kettlebell cleans before each press or not. Right, left, right . . .

5

GET-UP

Stability, mental ballet, balance, tension, total-body workout, strength endurance, and *control.* These are just some of the many words used to describe the get-up movement, and all are correct.

When I start teaching a new kettlebell student the get-up, they usually don't have very nice words to say about it. But in time, they will praise it as they start to see the benefits of doing it. It increases their stability and overall balance. They learn that they can get a total-body workout from doing something so primal and basic.

The get-up is a physical challenge of going from a horizontal position to a vertical position, standing up and going back down, all under control and with a weight above the person. In addition to that, one of the amazing things that I personally like about the get-up is the mental aspect of it. It can take me about 15 to 20 seconds to perform a set of 10 repetitions of kettlebell swings. However, it will take me about 1 to 2 minutes to complete 1 repetition on each side of get-ups. It's definitely a grind, but you're also performing a mental ballet as you think about the next step, and then the next one, and so on.

Another important aspect of the get-up is that you can break it down into sections and practice it. This especially helps if one of the sections is more challenging for you. Make that weak section one of your strengths by practicing it and practicing it. Then put it all together and you will find that your overall get-up strength and technique will improve.

One of my former kettlebell mentors was a strength coach for his local high school American football team. He taught them how to perform the swing and the get-up as their main training exercises for the upcoming season. They were taught a ballistic movement (swing) and a grind movement (get-up). Teaching them these two movements and their variations was what they needed to perform well on the football field. It was very simple yet effective.

Some of the benefits of training the get-up are improved shoulder stability and mobility, training the body as one unit, a stronger torso by increasing the reflexive stability of the core unit, and increased interaction of the various myofascial chains and intermuscular coordination. Other benefits include better communication between the brain and the motor units, developing patience and understanding of working through a grind movement while keeping the end goal in mind, and increasing your fat-burning ability.

Electromyography studies of the body tell us that the get-up generates over 100% peak activation of all four core muscles tested—rectus abdominis, external oblique, internal oblique, and transversus abdominis.

When I was playing semipro American football, the get-up was one of the exercises that I used on Tuesdays after my games on Saturday nights. It helped me to reconnect my whole body after playing for three or more hours as a defensive tackle. It also helped to make sure that I had the required thoracic mobility to perform any overhead movements, such as the kettlebell snatch and the barbell military press and overhead squat.

I teach my kettlebell students that once they learn the get-up, that they can use it in myriad ways. It can be used either without weight or with a light weight as part of their warm-up prior to doing their work sets, doing one to two repetitions per side. It can be performed as a part of cardio training, doing more repetitions (three to five) per side. Once you're ready for it, you can also perform the get-up as a strength exercise with heavier weight by doing one to two repetitions per side or as a get-up plus press, or merely just as a stand-alone exercise if you are short on time.

A word of caution and attention is needed here: spend a good amount of time to make sure that your quality is on point with this movement. Doing body-weight get-ups for several training sessions is prudent to your overall success with this exercise. Then go to the shoe get-up, and then add weight. Spend time at each section of the get-up and listen to your body. Make sure that every item is checked off your checklist at each section before moving on. You will be happy that you did this in the future; it will help you to avoid injury, improve strength, and teach you how to move gracefully through the entire sequence of movements.

Performing the get-up with attention to detail, using great technique, and with an iron mindset will pay dividends to you. The get-up is a great overall strength movement that can be used by anyone—ordinary people, professional athletes, grandparents—to help give them the strength and mobility needed for life or their sport.

EXERCISE FOCUS FOR KETTLEBELL GET-UPS

Each of the get-ups provides different ways to help you learn proper technique and develop strength. Here, we take a closer look at the focus of the exercises in this chapter.

Performing the get-up movement is one of the most basic and simple exercises in this book. With that being said, having quality technique is paramount to receiving the tremendous benefits while performing this exercise.

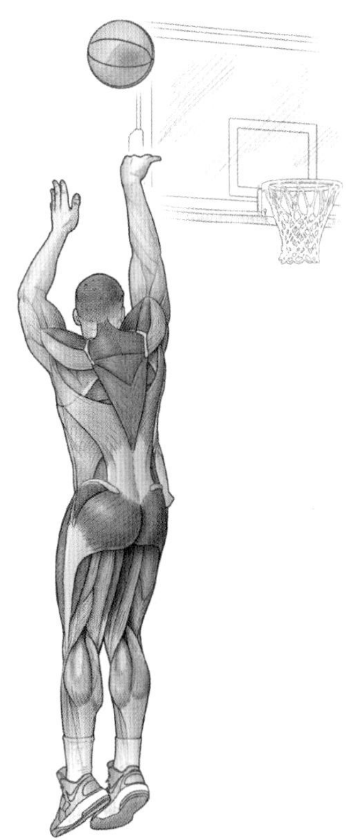

FIGURE 5.1 A basketball player needs to build upper-body endurance for shooting, blocking, rebounding, and passing.

The get-up improves the ability to reach above your head and hold a weighted object for a time, playing basketball (figure 5.1) or volleyball with your hands in the air, competing in track and field (especially the throwing events) and participating in gymnastics. Performing in these sports requires strong shoulders and upper back muscles to enable the athlete to perform at their highest ability while at the same time decreasing their injury potential.

At the top of the get-up, you will be performing a vertical plank position with a weight overhead, which will challenge your core section to maintain its rigidity while at the same time not allowing your lumbar spine to extend and your rib cage to flare open. As you learn this movement and are ready to progress from the body weight–only version, the addition of the shoe on your fist while performing the get-up adds a level of increased difficulty by preventing the shoe from falling off your fist while going from a horizontal position to a vertical position and back to a horizontal position. In some respects, this is more difficult than doing the get-up with a kettlebell as you are balancing a shoe while moving. By the time you progress to the kettlebell get-up, you should be well versed in the overall movement of the get-up, but now you are adding weight to it.

Remember, this is a grind exercise, so be judicious in adding extra weight. If you need more help with your kettlebell press (chapter 6), adding the press to each section of the get-up could enhance your overall pressing ability and further make your shoulders more resilient, especially if you compete in contact sports. This further improves your pressing ability from different angles while performing the get-up, providing a two-for-one benefit. Again, be mindful of your technique and emphasize quality over quantity.

Lastly, a variation of the get-up that will add to the already intense neuromuscular involvement of your core is to perform the get-up with the kettlebell in the bottom-up position (bell above the handle) throughout the entire movement.

BODY-WEIGHT (I.E., "NAKED") GET-UP   

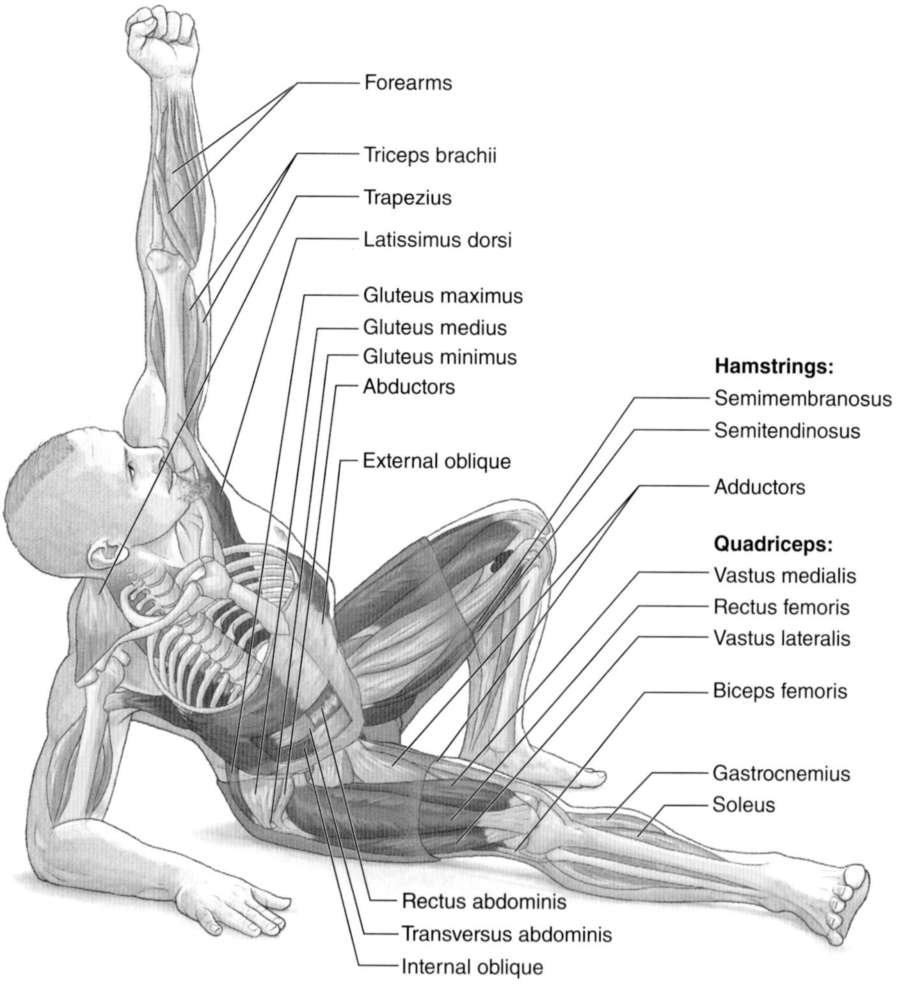

Forearms
Triceps brachii
Trapezius
Latissimus dorsi
Gluteus maximus
Gluteus medius
Gluteus minimus
Abductors
External oblique
Rectus abdominis
Transversus abdominis
Internal oblique

Hamstrings:
Semimembranosus
Semitendinosus
Adductors

Quadriceps:
Vastus medialis
Rectus femoris
Vastus lateralis
Biceps femoris
Gastrocnemius
Soleus

Execution

1. Lie on your back with your legs and arms at about 45-degree angles from your body.
2. Place your left fist in the air above your body with your arm straight, keeping your thumb pointing toward your head. Bend your left knee to approximately 90 degrees, making sure that your thigh is still at a 45-degree angle from the body and that your left foot is flat.
3. Roll up onto your right elbow by pressing through your left heel. Bring your left shoulder toward your right hip, like an approximately 45-degree sit-up. Pack your right shoulder; in other words, "suck the head" of your shoulder in and down toward your waist.
4. While keeping your eyes on your left fist, transition to pushing through the heel of the right hand by placing it on the ground and pointing the fingers 45 degrees or more behind you. Locking your right elbow, further pack your right shoulder by "corkscrewing" your humerus (externally rotating it), while not moving the right hand on the ground.
5. While loading your left heel and right hand, flex your right knee and bring it underneath you and lift your hips off the ground. Place your right knee in line with your right hand and your right shin perpendicular to your left leg and foot. Your right hand should be about a torso length away from you. At this point you will have three points of contact with the ground: your right hand, your right shin, and your left foot. Your left arm will be in a straight vertical line with your right arm.
6. Perform a slight hip hinge with your right hip, bringing your weight toward your right foot and left knee, and bring your torso to an upright position. Your left arm will be in an overhead position. Look forward, keeping your cervical spine neutral.
7. There are two ways to get into the half-kneeling lunge position:
    a. Move your right lower leg in a windshield-wiper fashion so that your right lower leg becomes parallel with your left leg and foot.
    b. Step your left leg and foot to the right to make it parallel with your right lower leg.
8. Before you stand, your left knee should be bent 90 degrees with the left foot flat and your right ankle dorsiflexed with your toes extended in a half-kneeling position. Your legs should be parallel to each other, like cross-country skis.
9. Stand up. You have now gone from a horizontal position to a vertical one.
10. Reverse the movements and positions that you performed to stand up and return to the horizontal start position. Control the descent, making sure that you don't flop back down to the ground.

(continued)

BODY-WEIGHT (I.E., "NAKED") GET-UP *(continued)*

Muscles Involved

Primary: Erector spinae (iliocostalis, longissimus, spinalis), quadriceps (rectus femoris, vastus lateralis, vastus medialis, vastus intermedius), gluteus maximus, hamstrings (semitendinosus, semimembranosus, biceps femoris), latissimus dorsi, rectus abdominis, external oblique, internal oblique, transversus abdominis

Secondary: Trapezius, rhomboids, gluteus medius, gluteus minimus, abductors and adductors, calf muscles (gastrocnemius, soleus), forearms (wrist flexors, finger flexors), elbow extensors (triceps brachii)

Anatomic Focus

Grip: The hand in the air will be closed and its wrist will be neutral.

Stance: When you are in the horizontal position, the feet should be squat-stance wide. When you are in the vertical position, the feet are hip-width to shoulder-width apart.

Trajectory: The body will go from a horizontal position to a vertical stance and back to horizontal.

Range of motion: The get-up is a whole-body exercise. It is a grind, not ballistic. When done properly, it will help make your shoulders bulletproof, make your core stronger than sit-ups will, and possibly add years to your life span. It can be done with body weight, a shoe on your fist, or a kettlebell or a dumbbell in your hand. Make sure that you master the technique of a body-weight get-up first before moving forward. Pause at each section. Pay attention to what your body is saying to you. If you need to move your foot or your hand before performing the next move, do so. Check that you are in the best position to succeed before performing it.

Technique note: Do not overextend your spine at the top of the get-up. Note the following when performing the get-down phase of the movement. Make sure that when you kneel down, you touch the ground softly with your knee. Place your hand on the ground about the same distance away from you as on the way up. Pack your shoulders before moving the down arm. When you get to your elbow position, with control, push your torso away from your elbow as you lie down. A still photograph taken at any portion of the get-up should not show whether you are going up or going down. Make both phases the same.

SHOE GET-UP

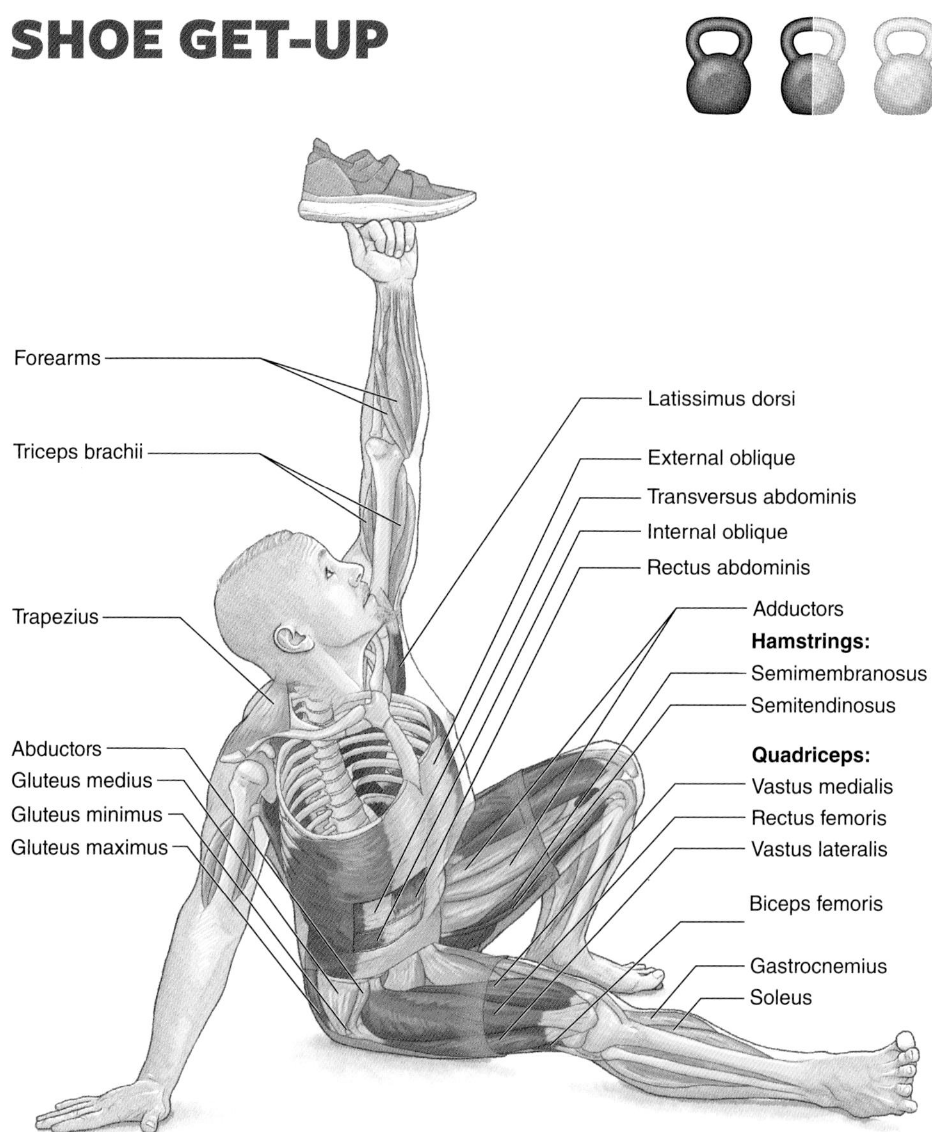

Execution

1. Lie on your back with your legs and arms at about 45-degree angles away from your body.
2. Place your left fist in the air above your body with your arm straight. Put a flat-bottomed shoe on top of your left fist with the proximal phalanges parallel to the ground.

3. Keep your thumb pointed toward your head. Bend your left knee to approximately 90 degrees, making sure that your thigh is still at a 45-degree angle from your body and that your left foot is flat.
4. Roll up onto your right elbow by pushing through your left heel and the right elbow, without moving your right elbow. Bring your left shoulder toward your right hip, doing a diagonal sit-up of approximately 45-degrees. Pack the head of your right shoulder in and down toward your waist.
5. While keeping your eyes on the shoe, transition to pushing through the heel of the right hand by placing it on the ground, pointing the fingers 45 degrees or more behind you. Locking your right elbow, further pack your right shoulder by corkscrewing your humerus externally, while not moving the right hand on the ground.
6. While loading your left heel and right hand, flex your right knee and bring it underneath you and lift your hips off the ground. Place your right knee in line with your right hand and your right shin perpendicular to your left leg and foot. Your right hand should be about a torso length away from you. At this point you will have three points of contact with the ground: your right hand, your right shin, and your left foot. Your left arm will be in a straight vertical line with your right arm.
7. Perform a slight hip hinge with your right hip, bringing your weight toward your right foot and left knee, and bring your torso to an upright position. Your left arm will be in an overhead position and your eyes will be looking forward with your cervical spine neutral.
8. There are two ways to get into the half-kneeling lunge position:
   a. Move your right lower leg in a windshield-wiper fashion so that your right lower leg becomes parallel with your left leg and foot.
   b. Step your left leg and foot to the right to make it parallel with your right lower leg.
9. Before you stand, your left knee should be bent 90 degrees with the left foot flat and your right ankle dorsiflexed with your toes extended in a half-kneeling position. Your legs should be parallel to each other, like cross-country skis.
10. Stand up. You have now gone from a horizontal position to a vertical one. It is now time to do the get-down.
11. Reverse the movements and positions that you performed to stand up while maintaining the shoe position on your fist and return to the horizontal start position. Control the descent, making sure that you don't flop back down to the ground.

(continued)

SHOE GET-UP *(continued)*

Muscles Involved

Primary: Erector spinae (iliocostalis, longissimus, spinalis), quadriceps (rectus femoris, vastus lateralis, vastus medialis, vastus intermedius), gluteus maximus, hamstrings (semitendinosus, semimembranosus, biceps femoris), latissimus dorsi, rectus abdominis, external oblique, internal oblique, transversus abdominis

Secondary: Trapezius, rhomboids, gluteus medius, gluteus minimus, abductors and adductors, calf muscles (gastrocnemius, soleus), forearms (wrist flexors, finger flexors), elbow extensors (triceps brachii)

Anatomic Focus

Grip: The hand in the air will be closed with the proximal phalanges parallel to the ground, holding a shoe on top. Your wrist will be neutral throughout the get-up.

Stance: When you are in the horizontal position, your feet should be squat-stance wide. When you are in the vertical position, your feet are hip-width to shoulder-width apart.

Trajectory: The body and the shoe will go from a horizontal position to a vertical stance and back to horizontal.

Range of motion: The get-up is a whole-body exercise. It is a grind, not ballistic. When done properly, it will help make your shoulders bulletproof, make your core stronger than sit-ups will, and possibly add years to your life span. It can be done with body weight, a shoe on your fist, or a kettlebell or a dumbbell in your hand. Make sure that you master the technique of a body-weight get-up first before moving forward. Pause at each section. Pay attention to what your body is saying to you. If you need to move your foot or your hand before performing the next move, do so. Check that you are in the best position to succeed before performing it.

Technique note: Do not overextend your spine at the top of the get-up. Note the following when performing the get-down phase of the movement. Make sure that when you kneel down, you touch the ground softly with your knee. Place your hand on the ground about the same distance away from you as on the way up. Pack your shoulders before moving the down arm. When you get to your elbow position, with control, push your torso away from your elbow as you lie down. A still photograph taken at any portion of the get-up should not show whether you are going up or going down. Make both phases the same.

Adding the shoe increases the challenge to you and your body as you keep the shoe on your fist from the ground to a standing position and back to the floor. You will find yourself moving through the get-up motion slower than you do in the body-weight or kettlebell versions of the exercise because of the shoe on your hand. If you drop the shoe at any point, my recommendation is to place the shoe back on top of your fist and continue from that point. This type of get-up is also useful to help you with your technique while at the same time giving you a challenge, while only using a shoe for weight.

VARIATION

Sandbag Get-Up

This variation is performed with a small sandbag, about four to six inches by four to six inches square (the size of a cornhole bag) on top of your hand. This is used instead of a shoe for those who have difficulty holding it in position because their fingers do not have a flat surface for the shoe.

KETTLEBELL GET-UP

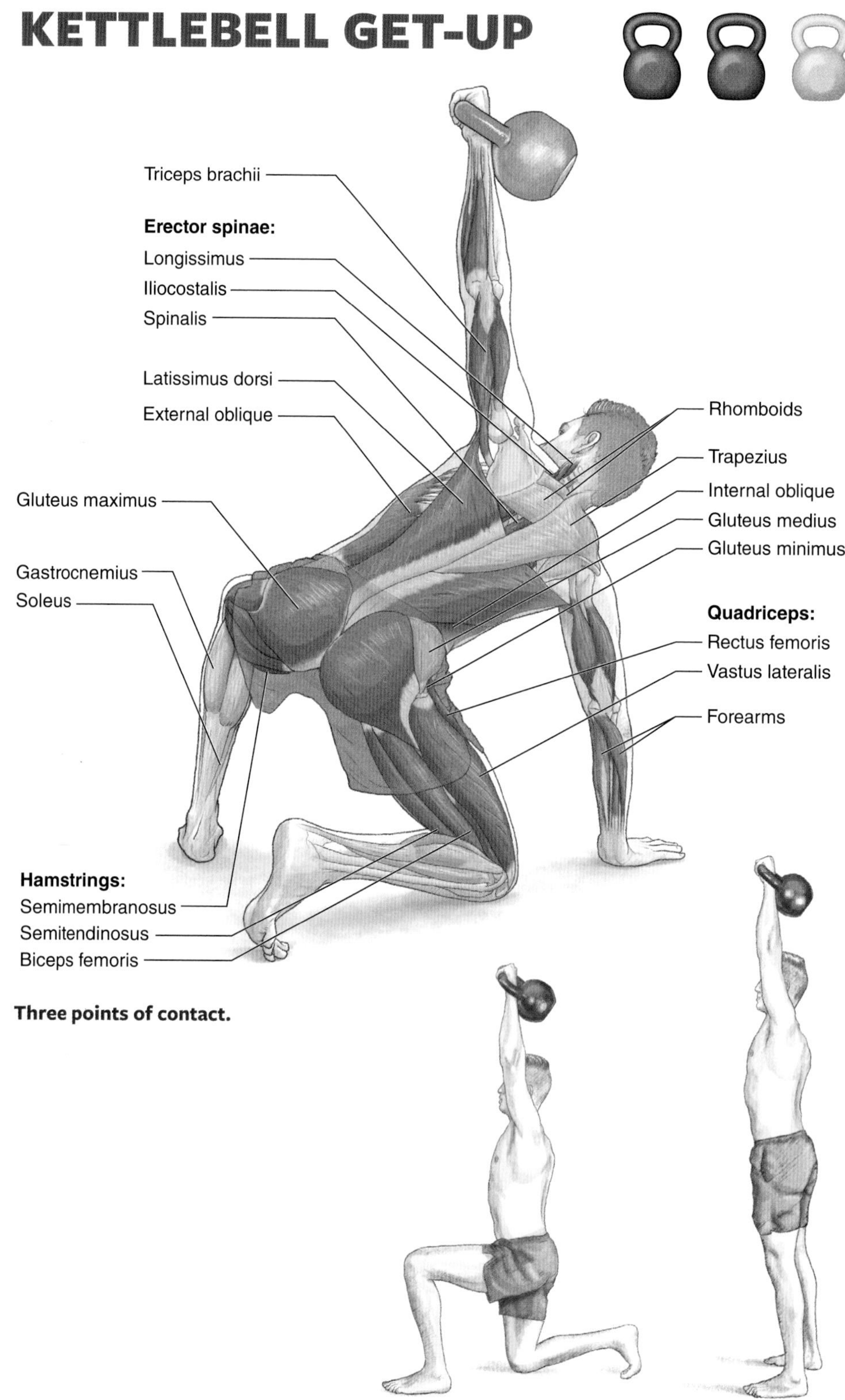

Three points of contact.

Half-kneeling position.

Stand up.

Execution

1. Lie on your left side with the kettlebell close to you. Grab the handle with your left hand and place your right hand with a thumbless grip over the top. While keeping the kettlebell close to you, roll to your back with the two-handed grip. The kettlebell will be down by your ribs with your left elbow on the ground.
2. Let go with your right hand and floor press the kettlebell up. You may also press the kettlebell up with both hands.
3. Keep your thumb pointed toward your head. Bend your left knee at approximately 90 degrees, making sure that your left leg is at a 45-degree angle to the right leg and that your left foot is flat. Angle your right leg 45 degrees from the left leg and move your arm to about a 45-degree angle from your torso.
4. Roll up onto your right elbow by pressing through your left heel and right elbow. Move into position without moving your right elbow. You will be rolling up, meaning bringing your left shoulder toward your right hip, doing an approximately 45-degree sit-up. Pack your right shoulder in and down toward your waist.
5. While keeping your eyes on the kettlebell, transition to pushing through the heel of the right hand by placing it on the ground, pointing the fingers 45 degrees or more behind you. Locking your right elbow, further pack your right shoulder by externally rotating your humerus, while not moving the right hand on the ground.
6. While loading your left heel and right hand, flex your right knee and bring it underneath you and lift your hips off the ground. Place your right knee in line with your right hand and your right shin perpendicular to your left leg and foot. Your right hand should be about a torso length away from you. At this point you will have three points of contact with the ground: your right hand, your right shin, and your left foot. Your left arm will be in a straight vertical line with your right arm.
7. Perform a slight hip hinge with your right hip, bringing your weight toward your right foot and left knee, and bring your torso to an upright position. Your left arm will be in a weighted overhead position and your eyes will be looking forward with your cervical spine neutral.
8. There are two ways to get into the half-kneeling lunge position:
   a. Move your right lower leg in a windshield-wiper fashion so that your right lower leg becomes parallel with your left leg and foot.
   b. Step your left leg and foot to the right to make it parallel with your right lower leg.
9. Before you stand, your left knee should be bent 90 degrees with the left foot flat and your right ankle dorsiflexed with your toes extended in a half-kneeling position. Your legs should be parallel to each other, like cross-country skis.

(continued)

KETTLEBELL GET-UP *(continued)*

10. Stand up. You have now gone from a horizontal position to a vertical one. It is now time to do the get-down.
11. Reverse the movements and positions that you performed to stand up and return to the horizontal start position. Control the descent, making sure that you don't flop back down to the ground.
12. There are two recommended ways to switch sides:
    a. With both hands, drag the kettlebell on the floor around your head in an arc to the other side.
    b. Move your body to the other side of the kettlebell.

Muscles Involved

Primary: Erector spinae (iliocostalis, longissimus, spinalis), quadriceps (rectus femoris, vastus lateralis, vastus medialis, vastus intermedius), gluteus maximus, hamstrings (semitendinosus, semimembranosus, biceps femoris), latissimus dorsi, rectus abdominis, external oblique, internal oblique, transversus abdominis, forearms (wrist flexors, finger flexors), elbow extensors (triceps brachii)

Secondary: Trapezius, rhomboids, gluteus medius, gluteus minimus, abductors and adductors, calf muscles (gastrocnemius, soleus)

Anatomic Focus

Grip: The kettlebell handle will be in a diagonal position from the web of the thumb to the pisiform of the hand. Your wrist will be neutral.

Stance: When you are in the horizontal position, your feet should be squat-stance wide. When you are in the vertical position, your feet are hip-width to shoulder-width apart.

Trajectory: The body and the kettlebell will go from a horizontal position to a vertical stance and back to horizontal.

Range of motion: The get-up is a whole-body exercise. It is a grind, not ballistic. When done properly, it will help make your shoulders bulletproof, make your core stronger than sit-ups will, and possibly add years to your life span. It can be done with body weight, a shoe on your fist, or a kettlebell or a dumbbell in your hand. Make sure that you master the technique of a body-weight get-up first before moving forward. Pause at each section.

Pay attention to what your body is saying to you. If you need to move your foot or your hand before performing the next move, do so. Check that you are in the best position to succeed before performing it.

Technique note: Do not overextend your spine at the top of the get-up. Note the following when performing the get-down phase of the movement. Make that sure when you kneel down, you touch the ground softly with your knee. Place your hand on the ground about the same distance away from you as on the way up. Pack your shoulders before moving the down arm. When you get to your elbow position, with control, push your torso away from your elbow as you lie down. A still photograph at any portion of the get-up should not show whether you are going up or going down. Make both phases the same.

VARIATION

Bottom-Up Get-Up

To perform this variation, hold the kettlebell by its handle with the body bottom up. You will need to crush-grip the handle and prevent the body of the kettlebell from falling during the get-up. This variation is only for advanced athletes. This variation also provides more challenge for your core muscles, more so than the regular get-up.

KETTLEBELL GET-UP PLUS PRESS

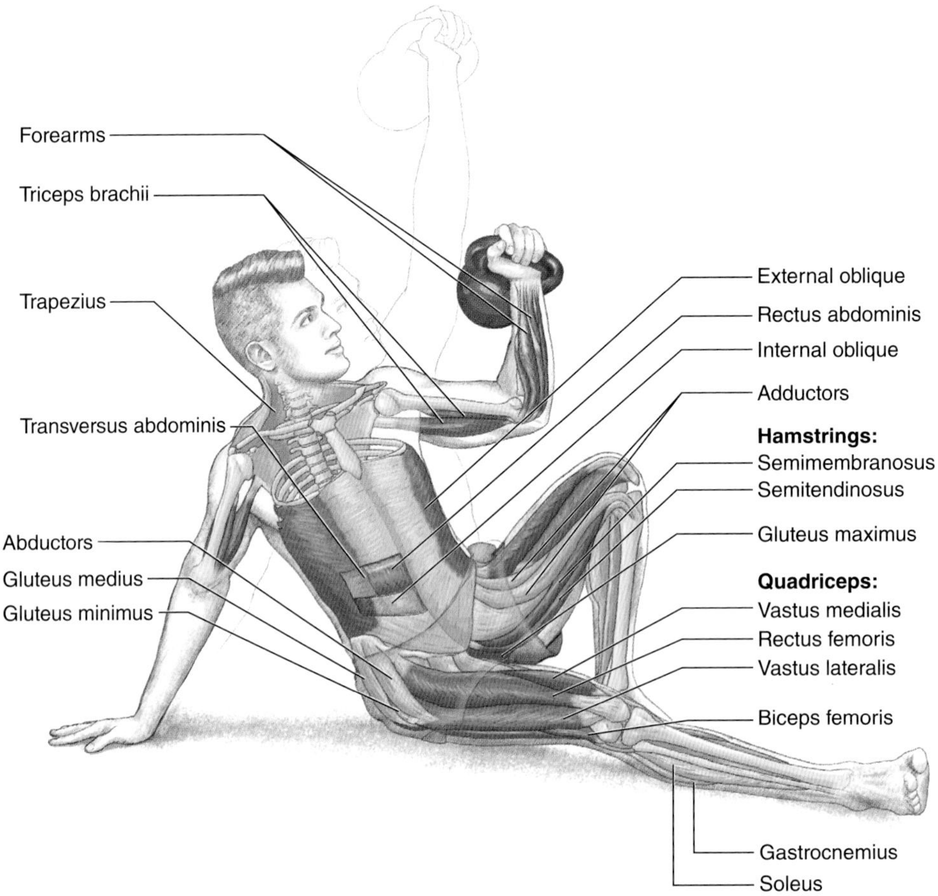

Forearms
Triceps brachii
Trapezius
Transversus abdominis
Abductors
Gluteus medius
Gluteus minimus
External oblique
Rectus abdominis
Internal oblique
Adductors
Hamstrings:
Semimembranosus
Semitendinosus
Gluteus maximus
Quadriceps:
Vastus medialis
Rectus femoris
Vastus lateralis
Biceps femoris
Gastrocnemius
Soleus

Execution

1. Lie on your left side with the kettlebell close to you. Grab the handle with your left hand and place your right hand with a thumbless grip over the top. While keeping the kettlebell close to you, roll to your back with the two-handed grip. The kettlebell will be down by your ribs with your left elbow on the ground.
2. Let go with your right hand and floor press the kettlebell up. Repeat the floor press.
3. Keep your thumb pointed toward your head. Bend your left knee to approximately 90 degrees, making sure that it is at a 45-degree angle to the right leg and that your left foot is flat. Angle your right leg at a

45-degree angle from the left leg and move your arm to about a 45-degree angle from your torso.

4. Roll up onto your right elbow by pressing through your left heel and right elbow, without moving your right elbow. You will be rolling up, meaning bringing your left shoulder toward your right hip, to get into an approximately 45-degree sit-up. Pack your right shoulder in and down toward your waist. Press the kettlebell.
5. While keeping your eyes on the kettlebell, transition to pushing through the heel of the right hand by placing it on the ground, pointing the fingers 45 degrees or more behind you. Locking your right elbow, further pack your right shoulder by externally rotating your humerus and not moving the right hand on the ground. Press the kettlebell.
6. While loading your left heel and right hand, flex your right knee and bring it underneath you and lift your hips off the ground. Place your right knee in line with your right hand and your right shin perpendicular to your left leg and foot. Your right hand should be about a torso length away from you. At this point you will have three points of contact with the ground: your right hand, your right shin, and your left foot. Your left arm will be in a straight vertical line with your right arm.
7. Perform a slight hip hinge with your right hip, bringing your weight toward your right foot and left knee, and bring your torso to an upright position. Your left arm will be in a weighted overhead position and your eyes will be looking forward with your cervical spine neutral.
8. There are two ways to get into the half-kneeling lunge position:
   a. Move your right lower leg in a windshield-wiper fashion so that your right lower leg becomes parallel with your left leg and foot.
   b. Step your left leg and foot to the right to make it parallel with your right lower leg.
9. Before you stand, your left knee should be bent 90 degrees with the left foot flat and your right ankle dorsiflexed with your toes extended in a half-kneeling position. Your legs should be parallel to each other, like cross-country skis. Perform an overhead press with the kettlebell.
10. Stand up. You have now gone from a horizontal position to a vertical one. Perform an overhead press with the kettlebell. It is now time to do the get-down.
11. Reverse the movements and positions that you performed to stand up and return to the horizontal start position. Control the descent, making sure that you don't flop back down to the ground.
12. There are two recommended ways to switch sides:
    a. With both hands, drag the kettlebell on the floor around your head in an arc to the other side.
    b. Move your body to the other side of the kettlebell.

(continued)

KETTLEBELL GET-UP PLUS PRESS *(continued)*

Muscles Involved

Primary: Erector spinae (iliocostalis, longissimus, spinalis), quadriceps (rectus femoris, vastus lateralis, vastus medialis, vastus intermedius), gluteus maximus, hamstrings (semitendinosus, semimembranosus, biceps femoris), latissimus dorsi, rectus abdominis, external oblique, internal oblique, transversus abdominis, forearms (wrist flexors, finger flexors), elbow extensors (triceps brachii)

Secondary: Trapezius, rhomboids, gluteus medius, gluteus minimus, abductors and adductors, calf muscles (gastrocnemius, soleus)

Anatomic Focus

Grip: Hold the kettlebell handle in a diagonal position from the web of the thumb to the pisiform of the hand. Your wrist will be neutral.

Stance: When you are in the horizontal position, your feet should be squat-stance wide. When you are in the vertical position, your feet should be hip-width to shoulder-width apart.

Trajectory: Move your body and the kettlebell from a horizontal position to a vertical stance and back to horizontal.

Range of motion: The get-up is a whole-body exercise. It is a grind, not ballistic. When done properly, it will help make your shoulders bulletproof, make your core stronger than sit-ups will, and possibly add years to your life span. This variation of the get-up, coupled with the kettlebell press, will turn an isometric contraction of the kettlebell shoulder into an isotonic contraction. In addition, this will further strengthen your pressing ability.

Technique note: Here are a few considerations:

1. Choose a lighter weight than what you can normally press.
2. Decide before you start the repetition which sections you wish to press at and how many repetitions for each section. If you press the kettlebell for one repetition at each section, you will press the kettlebell nine times while performing one repetition of the get-up.
3. Perform one to three repetitions per pressing section.
4. Most people press at the half-kneeling and standing sections. It is your choice.
5. Keep your get-up form while performing the pressing.
6. Your pressing shoulder will get fatigued. Pay attention to that and your technique. Emphasize quality over quantity.

Pay attention to what your body is saying to you. If you need to move your foot or your hand before performing the next move, do so. Check that you are in the best position to succeed before performing it.

Do not overextend your spine at the top of the get-up. Consider the following when performing the get-down phase of the movement. Make sure that when you are kneeling down, you touch the ground softly with your knee. Place your hand on the ground about the same distance away from you as on the way up. Pack your shoulders before moving the down arm. When you get to your elbow position, with control, push your torso away from your elbow as you lie down. A still photograph taken at any portion of the get-up should not show whether you are going up or going down. Make both phases the same.

6

SQUAT

Strong legs. Buns of steel. Flexibility. Lots of tension. Focused concentration. These are descriptive words describing the kettlebell squat. Having a strong lower body is important for athletics and everyday life and helps you to move about this world.

Before I go much further talking about the squat, let me establish some ground rules. The first is to keep your spine neutral, and the second is to keep your knees tracking your feet throughout the squat movement. One of the main things about any type of lifting, whether it be with a kettlebell, barbell, sandbag, or body weight, is that you keep your spine neutral when moving with load. That means your spine from the base of your skull to the tip of your tailbone must stay neutral when lifting.

With that being said, you should have enough flexibility and mobility that if you drop your toothbrush on the ground, you can bend over and pick it up any way you desire. You are not lifting a heavy object; therefore, you don't need to do any abdominal bracing or have a perfect set-up to pick up something as light as a toothbrush.

If you can't keep your spine neutral throughout the squat movement, though, what usually happens is your pelvis and hips will "butt wink" (A butt wink occurs near the bottom of a squat. The pelvis posteriorly tilts and causes the lumbar spine, especially the lower aspect, to go into spinal flexion when under load. This is potentially bad as the joints and discs of the lower lumbar spine could be at risk for injury.). What is usually happening is your L5 and S1 spinal vertebrae will start to move apart in the rear aspect while under load. This is not a good recipe to be following. After a few thousand repetitions of doing that, eventually the posterior aspect of the disc will weaken and you will have a problem. Bulging or herniation of the disc material can be painful and may require therapy to heal. If you are not able to squat below parallel (your hips are below your knees) without "butt winking," then stop at the position where butt winking doesn't occur and work that range of motion until you can go deeper.

As mentioned previously, the second ground rule is to keep your knees tracking your feet. Way too many people let their knees cave in (i.e., they are under valgus stress) either on the way down or the way up or both during the squat motion. Again, if you are lifting a light load or no load, you could be fine. But if you are lifting a heavy load, make sure it doesn't happen. To keep your knees in alignment, make sure that as you start to descend with the kettlebell, you push your knees out and keep pushing them out as you return to standing. This will train you to keep your knees over your toes—known as your knees tracking your feet (or toes).

With these two ground rules, hopefully you never have to hear from your doctor that squats are bad for your knees or back.

It is said that the deadlift is the concentric brother of the squat. This means the squat has an eccentric portion immediately followed by the concentric portion. Unless you are really working the negative or eccentric part of the deadlift, the deadlift is generally a concentric-only lift.

The squat tends to be more of a hypertrophy-type lift, whereas the deadlift won't put much muscle on you. One of the advantages of the squat is you can also work the eccentric part on the way down, meaning the active negative is a large part of setting up the ascent of the squat. Make sure that you are pulling yourself down, while pushing your knees out at the same time. You will arrive at the turning point ready to ascend and do it with lots of strength and power.

One of the questions I get all the time regarding kettlebell training is what kind of shoes to wear. My answer is either train barefoot or in a zero-rise, toe-to-heel shoe. Do not train in running shoes or an athletic shoe that has a thick sole. Using this will throw you off a solid foundation with which to do your lifting.

By performing the kettlebell squat properly and with intention, you will realize the huge benefits of squatting: strong legs, buns of steel, flexibility, sturdy abdominal muscles, and an iron will to stand up with some weight! Athletes in American football, basketball, tennis, track and field, gymnastics, soccer, powerlifting, strongman competitions, weightlifting, volleyball, and ice hockey (figure 6.1) benefit from effective squat training.

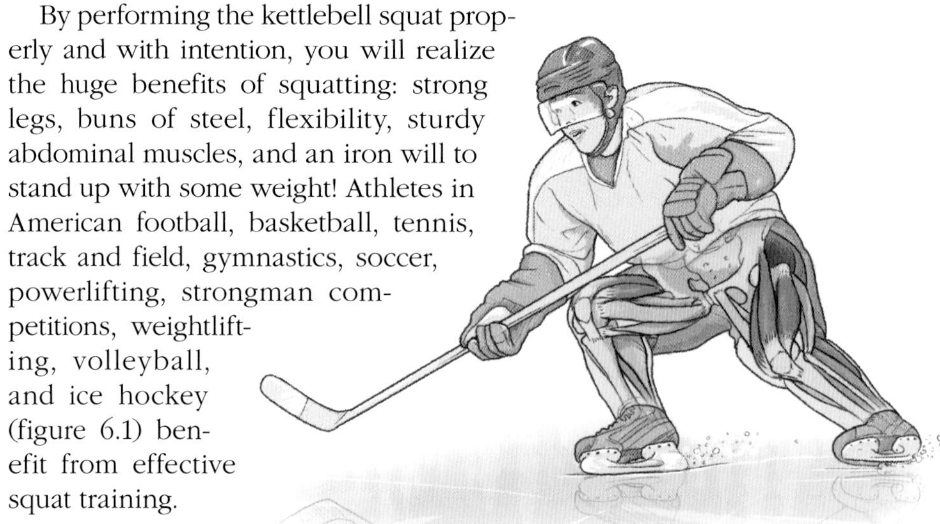

FIGURE 6.1 An ice hockey player understands the importance of a strong lower body.

EXERCISE FOCUS FOR KETTLEBELL SQUATS

Each category of squats provides different ways to help you learn proper technique and develop strength. Here, we take a closer look at the focus of the exercises in each category.

We have been squatting since we were infants, so we already have the basic skill. To further enhance this ability, perform the goblet squat to introduce squatting with load. The many benefits of performing goblet squats start with developing a strong squat, which is used in daily activities, running and jumping, and many sports, such as American football, basketball, soccer, volleyball, hockey, powerlifting, and weightlifting. In addition to that, being able to create and hold a vertical plank position momentarily at the top of the lift; learning proper tension-developing skills, which will be used later with other strength movements; and learning the neutral spine position and how to keep it throughout the movement are all paramount to be able to squat properly and efficiently. Keeping your knees tracking your toes will not only benefit you during the squat but also in other kettlebell movements. Doing this during any lower body movement will help decrease acute and chronic injury to the knee joint, and it will also keep the entire lower extremity linked up. This will then help increase your athletic performance. Performing the prying goblet squat, while not necessarily a strength move and considered more of a mobility move, will help you develop a mobile squat and will help you open up your hips and provide greater range of motion for the hip and pelvis structures, which is important for anyone involved in athletics.

The double and single kettlebell front squats will both help develop a strong squat, allow you to hold a vertical plank position at the top of the lift under heavier weight, and will build a strong lower body. In addition, the single kettlebell front squat will load you asymmetrically, thereby recruiting additional muscles on the nonloaded side, which include the contralateral gluteal muscles and quadratus lumborum. This will benefit those athletes who participate in asymmetrical sports, such as tennis, golf, baseball, soccer, and rock climbing.

The Cossack squat can be both a strength and a mobility movement, depending on the weight of the kettlebell that you are using. The hack squat is primarily a strength move. Both will help you increase neuromuscular control of your ankles, knees, and hips and build a more mobile and resilient lower body. The hack squat will also further develop the anterior lower body in addition to working on your balance. The hack squat helps you to

- develop a strong squat, which is used in daily activities and many sports to help you run and jump;
- develop resilient and strong ankles, knees, and hips while increasing your balance;

- create and hold a vertical plank position momentarily at the top of the lift;
- learn proper tension-developing skills which will be used with other strength movements; and
- learn the neutral spine position and how to keep it throughout the movement.

GOBLET SQUAT

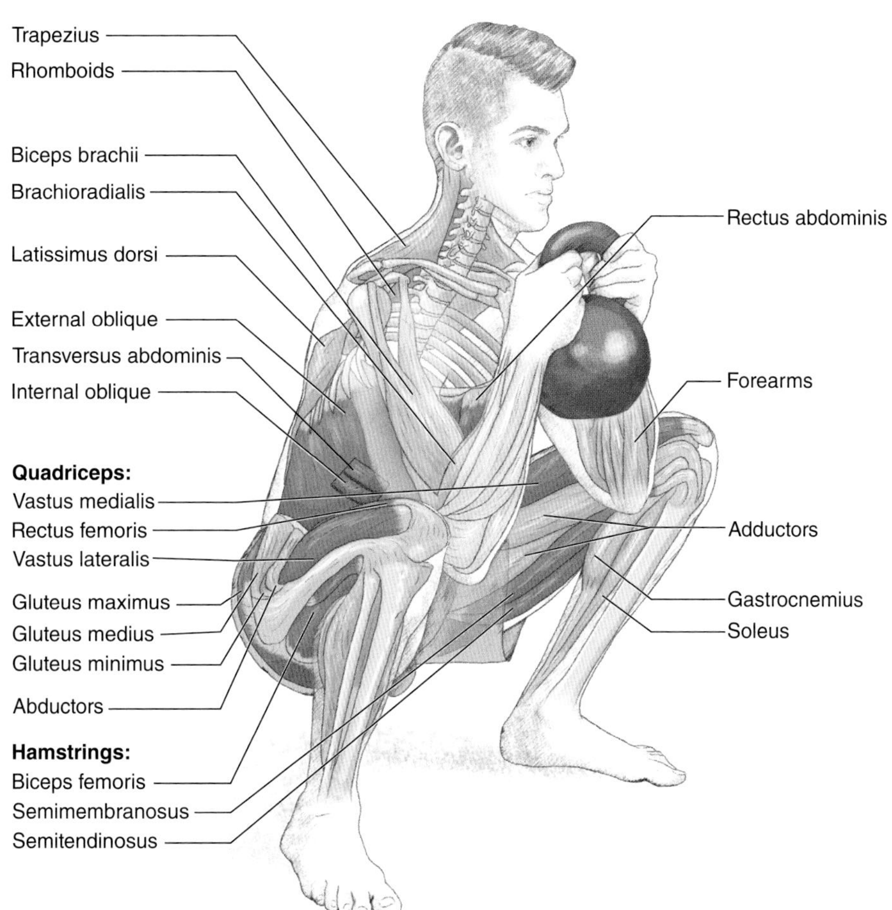

Execution

1. You have two options to get the kettlebell into position to perform the goblet squat:
   a. Place your feet about a foot (0.3 m) behind the kettlebell. Take a shoulder-width stance behind the kettlebell with the toes pointed slightly outward. Hike the kettlebell back with both hands on the handle (as in the swing exercises discussed in chapter 3) and stand up, keeping the kettlebell close to your body as you bend both elbows. You will end up holding the kettlebell with both hands, with one hand on each side.

(continued)

GOBLET SQUAT *(continued)*

b. Stand over the kettlebell in a shoulder-width stance with your toes pointed slightly outward (as in the deadlift exercises in chapter 2). Grip the handle with both hands. As you stand up to deadlift the kettlebell, continue the motion and bend your elbows so your hands end up being on each side of the handle at your chest level.

2. Prior to descending, point your toes slightly outward and tense your body (as in the kettlebell deadlift) at the top.
3. Start your descent by flexing your knees and hips equally. Descend until your hips are below your knees, keeping your spine neutral throughout the movement. Push your knees out so that your knees are tracking your toes both during the descent and the ascent.
4. While at the bottom of the squat, push your elbows against the insides of your knees. Be careful to keep your toes on the ground while making your spine as long as you can.
5. Once you are at the bottom, contract your abdominal muscles as hard as you can and ascend, keeping the abdominal muscles braced. Make sure that your hips and shoulders ascend at the same time.
6. At the top, assume the vertical plank position before starting the next repetition and repeat until you reach the desired number of repetitions.

Muscles Involved

Primary: Erector spinae (iliocostalis, longissimus, spinalis), quadriceps (rectus femoris, vastus lateralis, vastus medialis, vastus intermedius), gluteus maximus, hamstrings (semitendinosus, semimembranosus, biceps femoris), rectus abdominis, external oblique, internal oblique, transversus abdominis

Secondary: Trapezius, rhomboids, gluteus medius, gluteus minimus, abductors and adductors, latissimus dorsi, calf muscles (gastrocnemius, soleus), forearms (wrist flexors, finger flexors), elbow flexors (biceps brachii, brachioradialis)

Anatomic Focus

Hand spacing: Position your hands on opposite sides of the handle of the kettlebell.

Grip: Use a double overhand grip (supinate both hands at the top).

Stance: Position the legs approximately shoulder-width apart. Point your toes slightly outward.

Trajectory: Move the kettlebell in the vertical plane from start to finish.

Range of motion: At the beginning of each repetition, your toes will be rooted to the ground, your kneecaps pulled up, your gluteal muscles squeezed, and your abdominal muscles and latissimi dorsi engaged strongly. As in the kettlebell deadlift, the erector spinae, abdominal muscles, and latissimus dorsi muscles will help to stabilize and straighten the spine while the quadriceps, abductors and adductors of the hip, the gluteus maximus, and the hamstrings generate the squatting motion. Antishrug your shoulders, further activating your latissimus dorsi muscles. Your spine should be straight and stiff throughout the movement, even when your hips descend below your knees. If there is any butt winking, stop the squat motion before the wink happens and work through that range of motion. When your technique has improved, attempt a lower squat, again working on not butt winking. The lower you go below parallel, the greater the activation of the gluteus maximus. Do not overextend your spine at the top of the squat.

PRYING GOBLET SQUAT   

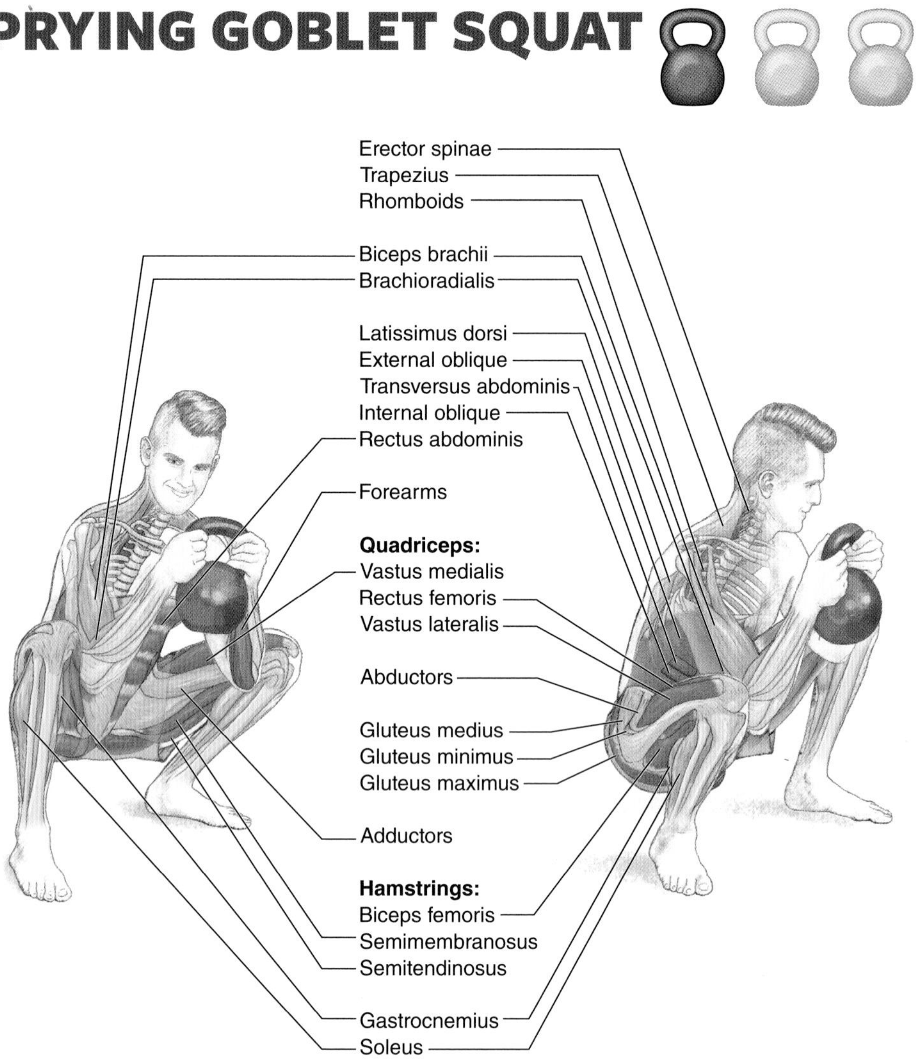

Execution

1. You have two options to get the kettlebell into position before performing the prying goblet squat:
    a. Place your feet about a foot (0.3 m) behind the kettlebell. Take a shoulder-width stance behind the kettlebell with your toes pointed slightly outward. Hike the kettlebell back with both hands on the handle (as in the swing exercises in chapter 3) and stand up, keeping the kettlebell close to your body as you bend both elbows. You will end up holding the kettlebell with both hands, one on each side.

b. Stand over the kettlebell in a shoulder-width stance with your toes pointed slightly outward (as in the deadlift exercises in chapter 2). Grip the handle with both hands. As you stand up to deadlift the kettlebell, continue the motion and bend your elbows so your hands end up being on each side of the handle at your chest level.

2. Prior to descending, turn your feet slightly out and tense your body (as in the kettlebell deadlift) at the top.
3. Start your descent by flexing your knees and hips equally. Descend until your hips are below your knees, keeping your spine neutral throughout the movement. Push your knees out so that your knees are tracking your toes both during the descent and the ascent.
4. When you are at the bottom of the squat, push your elbows against the inside of your knees. Be careful to keep your toes on the ground while making your spine as long as you can.
5. Let some of your air out but maintain your structural integrity.
6. While keeping your whole foot on the ground, move your hips side to side, going a little farther with each repetition. Try to separate the two halves of your pelvis. Move to each side two or three times, then stand up. Repeat for three to five repetitions for the first set.
7. An alternative is to move your pelvis in a figure 8 motion. First go one direction, then the opposite direction. Try to make bigger figure 8s each time, keeping your whole foot on the ground. Stand up. Repeat three to five repetitions.
8. At the top, assume the vertical plank position before starting the next repetition and repeat until you reach the desired number of repetitions.

Muscles Involved

Primary: Erector spinae (iliocostalis, longissimus, spinalis), quadriceps (rectus femoris, vastus lateralis, vastus medialis, vastus intermedius), gluteus maximus, hamstrings (semitendinosus, semimembranosus, biceps femoris), rectus abdominis, external oblique, internal oblique, transversus abdominis

Secondary: Trapezius, rhomboids, gluteus medius, gluteus minimus, abductors and adductors, latissimus dorsi, calf muscles (gastrocnemius, soleus), forearms (wrist flexors, finger flexors), elbow flexors (biceps brachii, brachioradialis)

(continued)

PRYING GOBLET SQUAT *(continued)*

Anatomic Focus

Hand spacing: Position the hands on opposite sides of the handle of the kettlebell.

Grip: Use a double overhand grip (supinate both hands at the top).

Stance: Position the legs approximately shoulder-width apart. Slightly point your toes outward.

Trajectory: Move the kettlebell in the vertical plane from the top to the bottom. At the bottom, the kettlebell will move with the hips, still being held by the hands.

Range of motion: At the beginning of each repetition, your toes will be rooted to the ground, your kneecaps pulled up, your gluteal muscles squeezed, and your abdominal and latissimus dorsi muscles strongly engaged. As in the kettlebell deadlift, the erector spinae, abdominal muscles, and latissimus dorsi muscles will help to stabilize and straighten the spine while the quadriceps, abductors and adductors of the hip, the gluteus maximus, and hamstrings generate the squatting motion. Antishrug your shoulders, further activating your latissimus dorsi muscles. Your spine should be straight and stiff throughout the movement, even when your hips descend below your knees.

This movement is a mobility movement, not a strength movement. Therefore, use a weight that is easy to moderate but not heavy. Keep the whole foot on the ground to make sure that this is a hip movement and not an ankle movement. You should be able to go deeper by the end of your repetitions, but not at the expense of your neutral spine. Do not overextend your spine at the top of the squat. Please see chapter 10, Mobility, regarding more information on the prying goblet squat.

SINGLE KETTLEBELL FRONT SQUAT

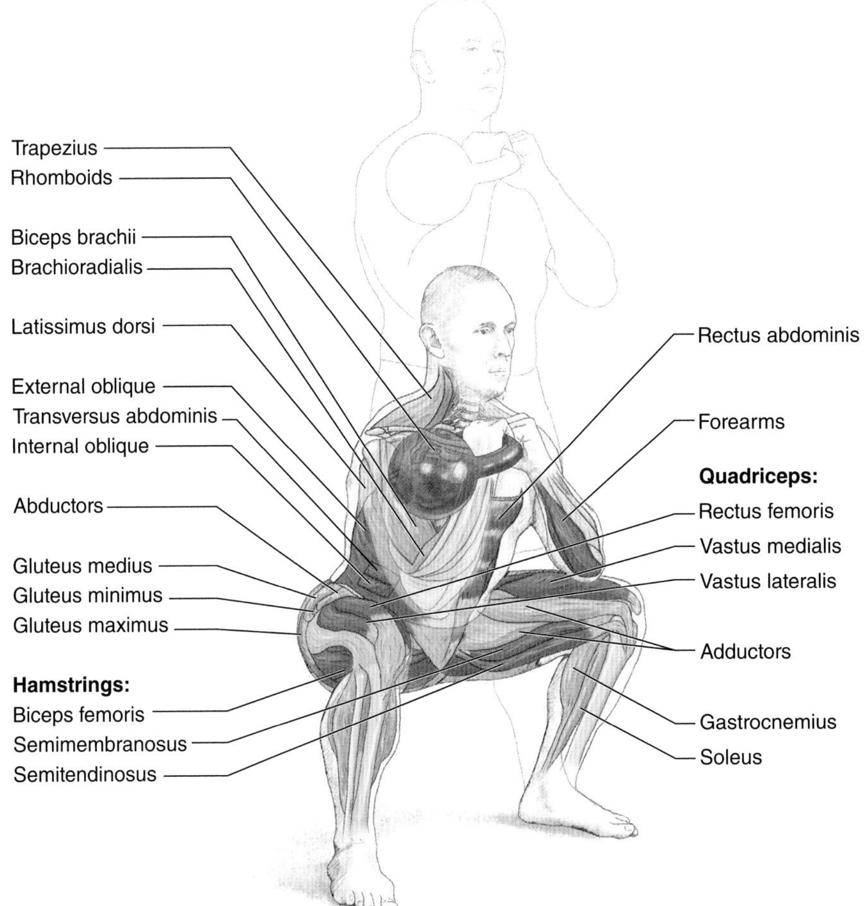

Execution

1. Clean a single kettlebell (see chapter 4 about how to clean a kettlebell) and hold it in the rack position, with your fingers around the handle and the bell resting on the top of your hand and front deltoid. Keep the wrist extended; don't let the weight of the kettlebell move your wrist out of position. Bring the other hand up and make a fist, holding this hand on the other side of the chest. Women: do not apply pressure to your breasts.
2. Prior to descending, point your feet slightly out and tense your body (as in the kettlebell deadlift exercises in chapter 2) at the top.
3. Start your descent by flexing your knees and hips equally. Descend until your hips are below your knees, keeping your spine neutral throughout the movement. Push your knees out so that your knees are tracking your toes during both the descent and the ascent.

(continued)

SINGLE KETTLEBELL FRONT SQUAT *(continued)*

4. Once at the bottom, contract your abdominal muscles as hard as you can and ascend to the top, keeping the abdominal brace. Make sure that your hips and shoulders ascend at the same time.
5. At the top, assume the vertical plank position before starting the next repetition and repeat until you reach the desired number of repetitions.

Muscles Involved

Primary: Erector spinae (iliocostalis, longissimus, spinalis), quadriceps (rectus femoris, vastus lateralis, vastus medialis, vastus intermedius), gluteus maximus, hamstrings (semitendinosus, semimembranosus, biceps femoris), rectus abdominis, external oblique, internal oblique, transversus abdominis

Secondary: Trapezius, rhomboids, gluteus medius, gluteus minimus, abductors and adductors, latissimus dorsi, calf muscles (gastrocnemius, soleus), forearms (wrist flexors, finger flexors), elbow flexors (biceps brachii, brachioradialis)

Anatomic Focus

Grip: Use a single overhand grip (position the hand in a neutral position).

Stance: Position the legs approximately shoulder-width apart. Slightly point your toes outward.

Trajectory: Move the kettlebell in the vertical plane from the top to the bottom.

Range of motion: At the beginning of each repetition, your toes will be rooted to the ground, your kneecaps pulled up, your gluteal muscles squeezed, and your abdominal and latissimus dorsi muscles strongly engaged. As in the kettlebell deadlift, the erector spinae, abdominal muscles, and latissimus dorsi muscles will help to stabilize and straighten the spine while the quadriceps, abductors and adductors of the hip, the gluteus maximus, and hamstrings generate the squatting motion. Antishrug your shoulders, further activating your latissimus dorsi muscles. Your spine should be straight and stiff throughout the movement, even when your hips descend below your knees.

Because you are using a single kettlebell during this entire movement, the load on the body will be asymmetrical. On the contralateral side, the muscles will be working harder to keep the body straight and the spine in a neutral position. Namely, the gluteal muscles and the quadratus lumborum on the nonweighted side are working hard, like they do in the single kettlebell swing exercises discussed in chapter 3. Do not overextend your spine at the top of the squat.

DOUBLE KETTLEBELL FRONT SQUAT

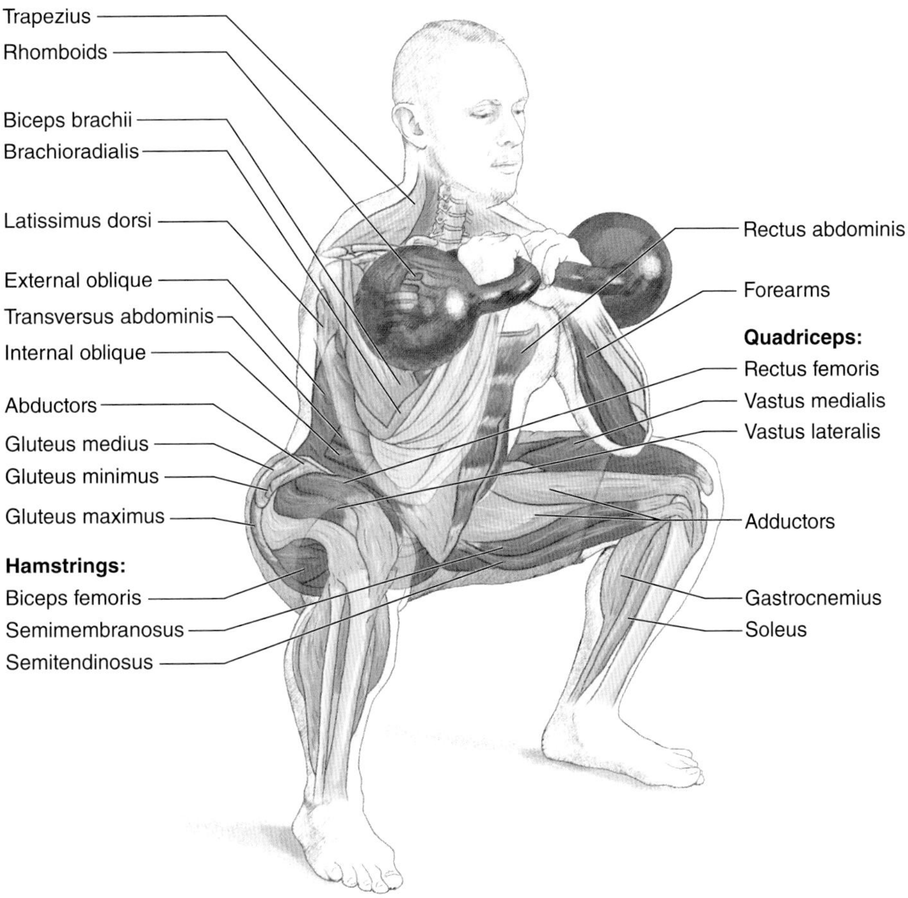

Execution

1. Do a clean lift of two kettlebells and hold them in the rack position, with your fingers around the handles and the bells resting on the tops of your hands and front deltoids. Keep the wrists extended; don't let the weight of the kettlebell move your wrists out of position. Men, you may interlock your fingers at the top. Women, do not apply pressure to your breasts.
2. Prior to descending, point your toes slightly outward and tense your body as in the kettlebell deadlift (chapter 2) at the top.

(continued)

DOUBLE KETTLEBELL FRONT SQUAT *(continued)*

3. Start your descent by flexing your knees and hips equally. Descend until your hips are below your knees, keeping your spine neutral throughout the movement. Push your knees out so that your knees are tracking your toes both during the descent and the ascent.
4. Once you are at the bottom, contract your abdominal muscles as hard as you can and ascend to the top, keeping the abdominal brace. Make sure that your hips and shoulders ascend at the same time.
5. At the top assume the vertical plank position before starting the next repetition and repeat until you reach the desired number of repetitions.

Muscles Involved

Primary: Erector spinae (iliocostalis, longissimus, spinalis), quadriceps (rectus femoris, vastus lateralis, vastus medialis, vastus intermedius), gluteus maximus, hamstrings (semitendinosus, semimembranosus, biceps femoris), rectus abdominis, external oblique, internal oblique, transversus abdominis

Secondary: Trapezius, rhomboids, gluteus medius, gluteus minimus, abductors and adductors, latissimus dorsi, calf muscles (gastrocnemius, soleus), forearms (wrist flexors, finger flexors), elbow flexors (biceps brachii, brachioradialis)

Anatomic Focus

Grip: Use a double overhand grip (position both hands neutral and facing each other).

Stance: Position the legs approximately shoulder-width apart. Slightly point your toes outward.

Trajectory: Move the kettlebell in the vertical plane from the top to the bottom.

Range of motion: At the beginning of each repetition, your toes will be rooted to the ground, your kneecaps pulled up, your gluteal muscles squeezed, and your abdominal and latissimus dorsi muscles strongly engaged. As in the kettlebell deadlift, the erector spinae, abdominal muscles, and latissimus dorsi muscles will help to stabilize and straighten the spine while the quadriceps, abductors and adductors of the hip, the gluteus maximus, and the hamstrings generate the squatting motion. Antishrug your shoulders, further activating your latissimus dorsi muscles. Your spine should be straight and stiff throughout the movement, even when your hips descend below your knees.

Keep both kettlebells on your chest; don't lift your elbows up. Do not overextend your spine at the top of the squat.

COSSACK SQUAT

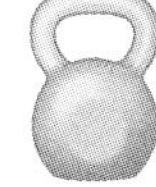

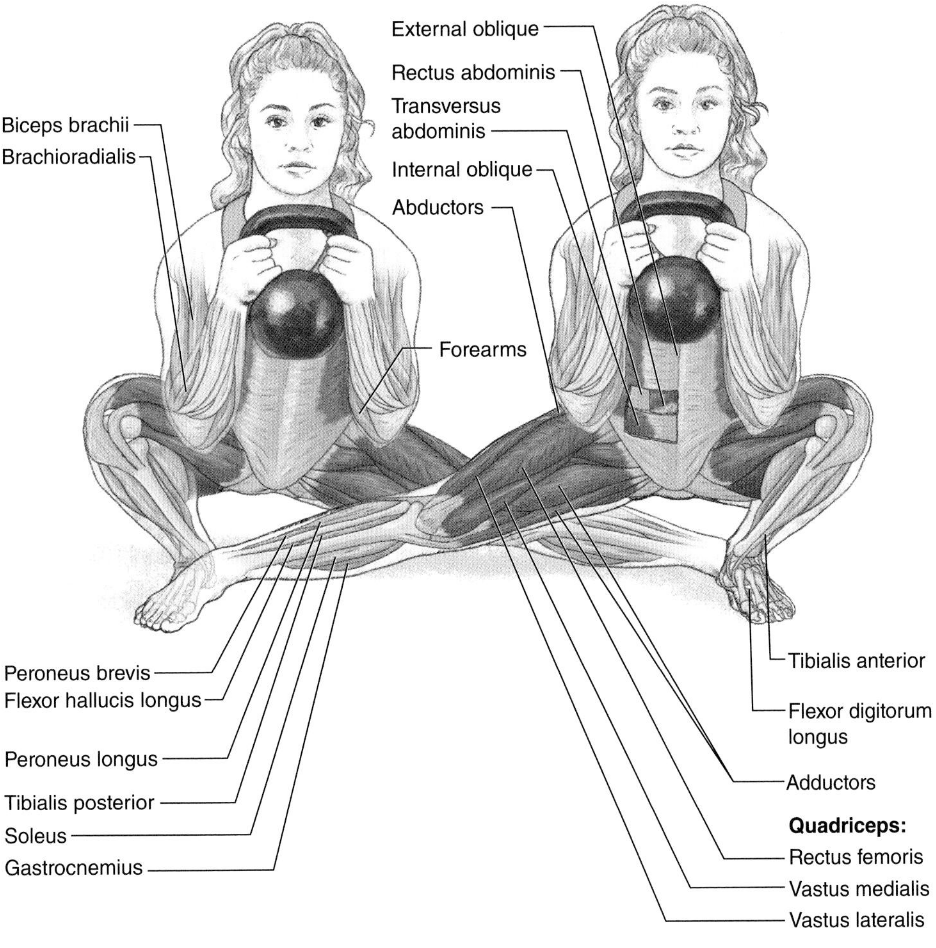

External oblique
Rectus abdominis
Transversus abdominis
Biceps brachii
Brachioradialis
Internal oblique
Abductors
Forearms
Peroneus brevis
Flexor hallucis longus
Peroneus longus
Tibialis posterior
Soleus
Gastrocnemius
Tibialis anterior
Flexor digitorum longus
Adductors
Quadriceps:
Rectus femoris
Vastus medialis
Vastus lateralis

Execution

1. You have two options to get the kettlebell into position to perform the Cossack squat:
   a. Place your feet about a foot (0.3 m) behind the kettlebell. Take a shoulder-width stance behind the kettlebell with your toes pointed slightly outward. Hike the kettlebell back with both hands on the handle (as in the swing exercises discussed in chapter 3) and stand up, keeping the kettlebell close to your body as you bend both elbows. You will end up holding the kettlebell with both hands, one on each side of the handle.

(continued)

COSSACK SQUAT *(continued)*

   b. Stand over the kettlebell in a shoulder-width stance with your toes pointed slightly outward (as in the deadlift exercises in chapter 2). Grip the handle with both hands. As you stand up to deadlift the kettlebell, continue the motion and bend your elbows so your hands end up being on each side of the handle at your chest level.

2. Move your feet to wider than shoulder width apart and slightly point your feet out. If needed, you can adjust your stance width in the bottom of the movement.
3. Laterally transfer your weight toward your right foot while keeping your left foot on the ground. Keep your knees tracking your toes while descending your hips below your knees, if possible. Make sure that your right foot is entirely planted and your left leg is straight. Keep your chest high while holding the kettlebell. An alternative is to allow your left toes to come off the ground, pointing toward the ceiling.
4. Shift your weight toward your left leg. There are two ways to do this: either stand up, achieving the vertical plank position before descending for the next repetition, or partially come up to allow your body to move to the other side. Make sure that your hips and shoulders ascend at the same time.
5. When to move from one leg to the other is based on your individual needs and goals. If your goal is to develop strength, stay on the right leg for a set time, such as three seconds, before moving to the left leg. If your goal is to develop mobility and increase cardio output, move from one leg to the other with no pause in between.

Muscles Involved

Primary: Erector spinae (iliocostalis, longissimus, spinalis), quadriceps (rectus femoris, vastus lateralis, vastus medialis, vastus intermedius), gluteus maximus, gluteus medius, gluteus minimus, abductors and adductors, hamstrings (semitendinosus, semimembranosus, biceps femoris), rectus abdominis, external oblique, internal oblique, transversus abdominis

Secondary: Trapezius, rhomboids, latissimus dorsi, quadratus lumborum, gastrocnemius, soleus, flexor hallucis longus, flexor digitorum longus, tibialis posterior, peroneus longus and brevis, tibialis anterior, forearms (wrist flexors, finger flexors), elbow flexors (biceps brachii, brachioradialis)

Anatomic Focus

Hand spacing: Position your hands on opposite sides of the handle of the kettlebell.

Grip: Use a double overhand grip (supinate your hands at the top).

Stance: Position your legs more than shoulder-width apart. Slightly point your toes outward.

Trajectory: Move the kettlebell in the vertical plane from the top to the bottom.

Range of motion: At the beginning of each repetition, your toes will be rooted to the ground, your kneecaps pulled up, your gluteal muscles squeezed, and your abdominal and latissimus dorsi muscles strongly engaged. As in the kettlebell deadlift, the erector spinae, abdominals, and latissimus dorsi muscles will help to stabilize and straighten the spine while the quadriceps, abductors and adductors of the hip, the gluteal muscles, and the hamstrings generate the squatting motion. Antishrug your shoulders, further activating your latissimus dorsi muscles. Your spine should be straight and stiff throughout the movement, even when your hips descend below your knees.

Because you are using a single kettlebell during this entire movement, the load on the body will be asymmetrical. When you keep the straight-leg foot on the ground throughout the movement, you will tend to feel it more in your adductors. When the toes are off the ground and pointing up, you will feel it more in the hamstrings of the straight leg.

Do not use heavy loads when performing this movement. Because the exercise incorporates a small amount of balance training as part of the exercise, a heavy load could potentially cause an injury or other compensatory mechanisms with negative consequences.

Technique note: This movement can be used as a mobility exercise prior to heavier lifting or intense exercise or can be added as a strength move. One of its many benefits is increasing your neuromuscular control of your lower limbs, which can help decrease or prevent injury in athletics or general life.

HACK SQUAT

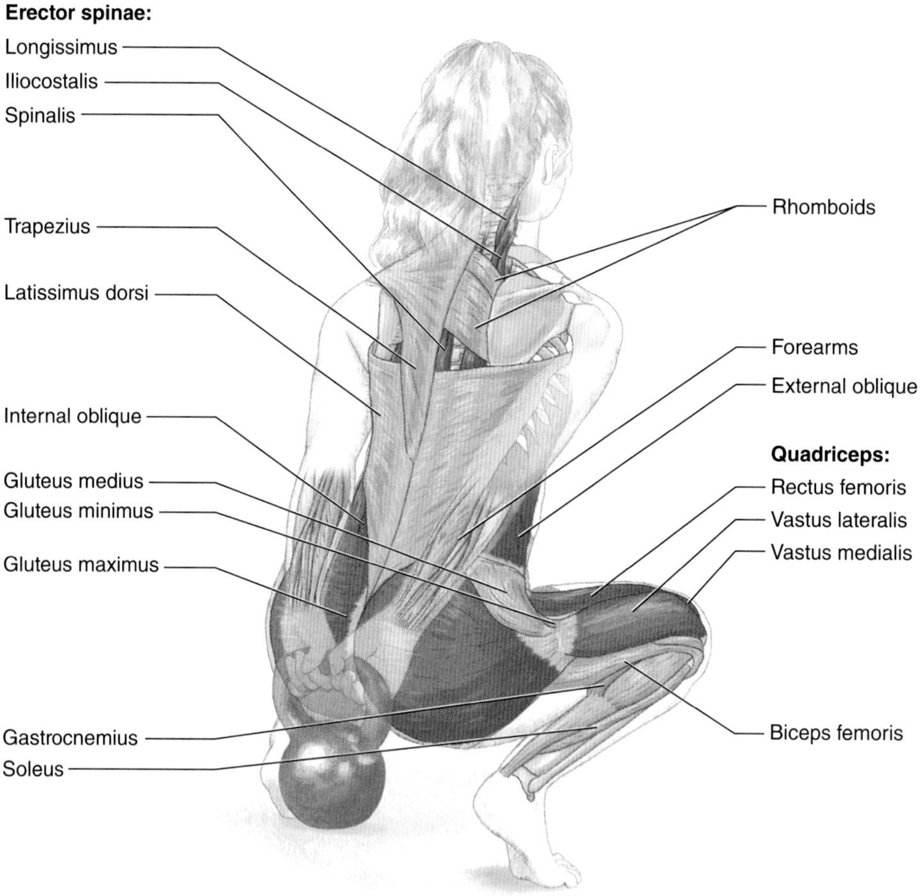

Execution

1. Stand over the kettlebell with a shoulder-width stance and your toes slightly pointed outward (as in the deadlift exercises in chapter 2). Grip the handle with both hands and deadlift it.
2. With one hand, move the kettlebell behind your back. Grip it with both hands. Flex your elbows slightly to place the kettlebell by your tailbone.
3. Place your heels close together and turn your toes 30 to 45 degrees out. Lift and open your chest and squeeze your shoulder blades together. Squeeze your gluteal muscles hard.

4. Keeping your torso tall and without leaning forward, pull your hips straight down by flexing your knees. Go as low as possible, keeping your spine neutral throughout the movement. Push your knees out so that your knees are tracking your toes both during the descent and the ascent. Your heels will come off the ground; that is normal.
5. Once you are at the bottom, contract your abdominal muscles as hard as you can and ascend to the top, keeping the abdominal brace.
6. At the top, assume the vertical plank position before starting the next repetition and repeat for the desired number of repetitions.

Muscles Involved

Primary: Erector spinae (iliocostalis, longissimus, spinalis), quadriceps (rectus femoris, vastus lateralis, vastus medialis, vastus intermedius), gluteus maximus, rectus abdominis, external oblique, internal oblique, transversus abdominis

Secondary: Trapezius, rhomboids, gluteus medius, gluteus minimus, abductors and adductors, latissimus dorsi, hamstrings (semitendinosus, semimembranosus, biceps femoris), calf muscles (gastrocnemius, soleus), forearms (wrist flexors, finger flexors)

Anatomic Focus

Hand spacing: Position your hands next to each other on the handle of the kettlebell.

Grip: Use a double overhand grip (pronate both hands at the top).

Stance: Stand with the heels touching or almost touching. Point the toes at an angle of 30 to 45 degrees outward.

Trajectory: Move the kettlebell in the vertical plane from start to finish.

Range of motion: At the beginning of each repetition, your toes will be rooted to the ground and turned out 30 to 45 degrees, your kneecaps pulled up, your gluteal muscles squeezed, and your abdominal and latissimus dorsi muscles strongly engaged. As in the kettlebell deadlift, the erector spinae, abdominal muscles, and latissimus dorsi muscles will help to stabilize and straighten the spine while the quadriceps, abductors and adductors of the hip, and the gluteus maximus generate the hack squatting motion. Antishrug your shoulders, further activating your latissimus dorsi muscles. Your spine should be straight and stiff throughout the movement, even when your hips descend below your knees.

(continued)

HACK SQUAT *(continued)*

You will go as low as possible, which is different for each person. This movement isn't for everyone; those with previous injuries or overuse wear and tear would be wise to steer clear of this exercise, at least in the beginning. Doing some body-weight hack squats, regular squats, or goblet squats would be more prudent until later. Do not overextend your spine at the top of the squat.

Technique note: Keeping your toe angle at 45 degrees increases the angle between your thighs, which puts more stress on your adductors and less on your quadriceps, whereas keeping it near 30 degrees puts more stress on your quadriceps. Regardless of toe angle, this type of squat will stress all your lower body muscles, especially the anterior muscles.

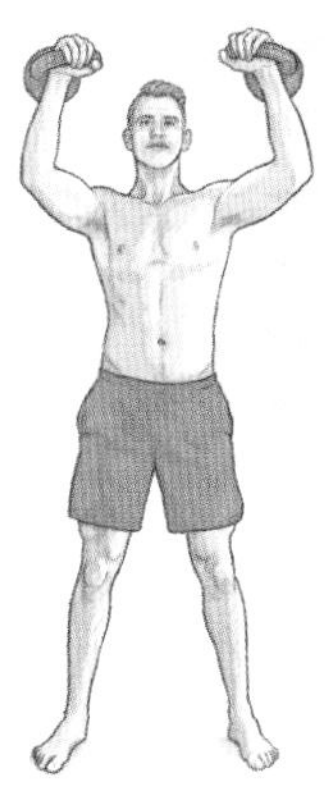

7 SNATCH

Swing. Snap (of the hips). *Punch. Strength endurance builder. Atomic. Impressive cardio. Brutal.* These are strong words said to me by my kettlebell students to describe the kettlebell snatch. The snatch is basically a summation of the skills learned from the earlier chapters in this book.

One of the benefits to becoming proficient with the snatch is improved cardiorespiratory endurance. You are putting a kettlebell overhead in a ballistic way. Unlike the kettlebell press, which is considered a grind movement with full-body tension, the snatch goes from tension to relaxation back to tension and so on. Decreased resting heart rate, improved blood pressure, reduced body fat and weight, and improved $\dot{V}O_2max$ are biomarkers of your body that will see improvements due to performing the kettlebell snatch.

As I alluded to in the swing chapter (chapter 3), your nervous system will improve its coordination and ability to maintain proper posture while moving the kettlebell, whether in a grind exercise or with the ballistic snatch. Your muscular system will also increase in strength and resiliency, which you will see in your everyday life and on the field or court of play.

When you start doing more and more snatches, you will find that the exercise could turn from an aerobic one to an anaerobic one, meaning you will go into oxygen debt. Do this slowly and build up to it. Just as I wouldn't recommend someone going out and running 10 miles for their first run, I strongly recommend that you take your time and get used to the snatch. Have some patience and enjoy the time it takes to develop the snatch skill. This is a total-body workout. Treat it as such.

Learn to control and enhance your breathing during the snatch. One of the ways to do this is via learning biomechanical breathing match. This also applies to the swing and the clean movements. Basically, it means that on the hike pass, breathe in. On the upswing, let a small amount of air out, behind tensed abdominal muscles, when your hips "snap" and your kneecaps pull up, regardless of whether you are doing the swing, clean, or snatch. If you are doing the snatch, this small expiration will occur before the kettlebell has reached the top position. Remember this: The snatch is a ballistic movement. Hence, your arm isn't lifting it, it's just merely escorting the kettlebell into place.

As you learn to perform the snatch with greater skill, push yourself on how long you can go. Can you do 10 repetitions per side for a total of 20 repetitions before setting the kettlebell down? Or 40 repetitions total? Maybe 100 repetitions? By controlling your breathing and continuing to acquire better snatch technique skills, you will end up burning more calories and enhancing your overall performance.

Let me give you friendly but strong words of caution with the snatch: Make sure that you can safely, and with good technique, put something overhead. If you can't, work on it and do something else until you can. Your body will thank you.

One of the stigmas about performing lots of kettlebell swings and snatches is what it could do to your hands. Good hand care is important here, and I'm not talking about wearing gloves to protect them. Go online and search "kettlebell hand care" and read about it. Go to the local pharmacy or buy products online to help you keep your well-earned calluses trimmed and well kept. If you don't watch after them, they could potentially stop you in your tracks as one or more of them could tear on you. This is not a fun way to end a training session.

Unlike the crush grip required for the kettlebell press, especially the bottom-up variety, the grip during the swing and the snatch is a looser grip. It is enough of a grip to control the kettlebell and not let it fly out of your hand. The kettlebell handle needs to move during the snatch, and if you are crush gripping the handle, that won't happen.

A benefit of performing the kettlebell snatch is that it helps to develop tenacious mental strength. You may need to shut your mind off while doing a prescribed number of snatches, as sometimes self-doubt creeps in and prematurely shuts the set down. The self-confidence that doing snatches develops is amazing to watch and see.

Spend some time learning how to perform the kettlebell snatch by first doing the exercises in earlier chapters. Then I recommend that you progress to snatching the kettlebell. At first, do a few repetitions per side then eventually move on to more repetitions. The kettlebell snatch is a tremendous exercise to learn and perform. You will be pleasantly surprised as to the positive changes that occur in your body and your mind.

EXERCISE FOCUS FOR KETTLEBELL SNATCHES

Each of the kettlebell snatches provides different ways to help you learn proper technique and develop strength. Here, we take a closer look at the focus of the kettlebell snatch.

When performing the kettlebell swing you are projecting the force you are generating in a horizontal fashion. The kettlebell snatch, while beginning as a kettlebell swing, is a vertical projection of force. One of the reasons we start this section of the kettlebell snatches with a special version of the kettlebell swing, namely the *Tyrannosaurus rex* (*T. rex*) swing, is because as we move from a horizontal projection of force to a vertical one, you need to learn what to do to decrease the overall distance from your body that the kettlebell travels during the kettlebell snatches. I learned in high school geometry that the shortest distance between two points is a straight line, and because the kettlebell snatch is a vertical projection of force, the closer you keep the kettlebell to your body, the less work you will need to do to complete a repetition. Hence, you will learn how to do the T. rex swing first prior to performing the actual snatch. Please remember that the T. rex swing is designed only to be used for the kettlebell snatch and is not recommended as a way to do a traditional kettlebell swing.

One of the benefits of doing the kettlebell T. rex swing is you will be keeping the kettlebell close to your body by bending the elbow to 90 degrees on the upswing, which will get you ready to learn and perform the kettlebell snatch. Just as with the one-hand swing, you will also learn to keep your shoulders and hips squared up throughout this movement even though you are using an asymmetrical load. You will recruit additional muscles, namely the contralateral gluteus medius, gluteus minimus, and quadratus lumborum, which will be further activated performing this type of snatch. Creating and holding a vertical plank position momentarily at the top of the lift, especially with the kettlebell going above your head in a ballistic way, you will be learning proper tension-developing skills.

The single kettlebell snatch helps you to develop a strong posterior chain, which is used in daily activities and many sports to help you run and jump. Sports in which your arm goes overhead, especially those where it is a sudden or ballistic move, would profit greatly from performing the kettlebell snatch. Examples of those that would benefit are weightlifting, basketball, volleyball, and swimming (figure 7.1).

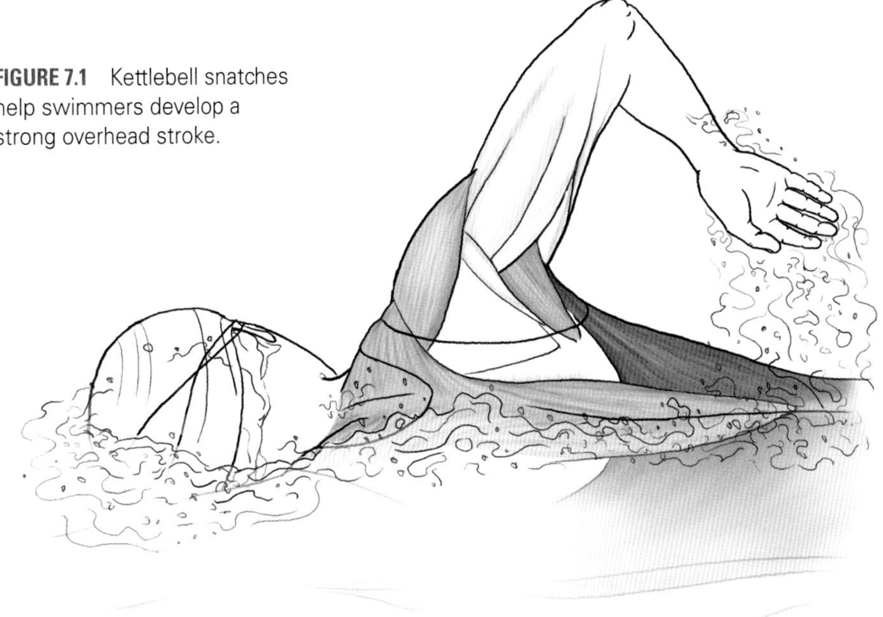

FIGURE 7.1 Kettlebell snatches help swimmers develop a strong overhead stroke.

Added benefits include developing better cardiorespiratory endurance and enhancing your body's ability to burn calories in a short time, developing greater mental strength and grit performing snatches and developing and producing more power to get the kettlebell above your head in a continuous motion. In addition, the double kettlebell snatch helps you to further enhance your cardiorespiratory endurance and enhance your body's ability to burn calories in a shorter time using heavier combined weight. Lastly, the single kettlebell snatch to eccentric press helps you to build an overall stronger press and increase upper-body strength, while also getting the benefits of the snatch.

SINGLE KETTLEBELL T. REX SWING WITH ONE HAND

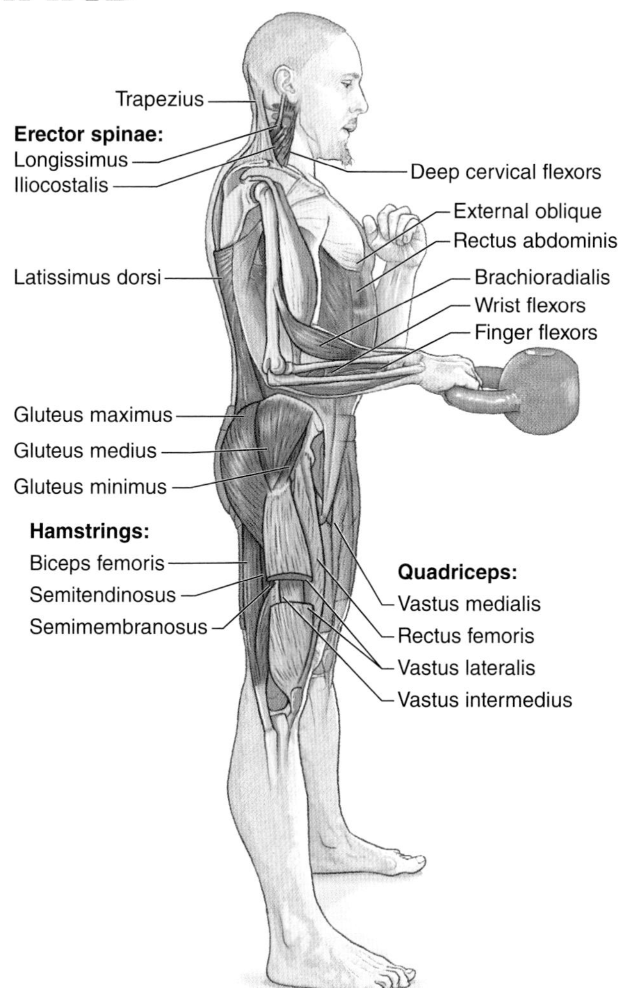

NOTE: This exercise prepares you for the snatch movements in the rest of the chapter.

Execution

1. Place your feet about a foot (0.3 m) behind the kettlebell. Take a shoulder-width stance behind the kettlebell with your toes pointed slightly outward.
2. Keeping your hips above your knees and below your shoulders, grip the kettlebell handle with one hand, placing your body into a hip-hinge position. You should feel tension in your hamstrings. Tilt the handle toward you, allowing it to become an extension of your arm.

(continued)

SINGLE KETTLEBELL T. REX SWING WITH ONE HAND *(continued)*

3. Place your nonworking hand beside the kettlebell but not on it. Doing this will help keep your shoulders square at the beginning.
4. Grip the ground with your toes, tighten the latissimi dorsi, abdominal muscles, and grip, and hike pass the kettlebell back between your legs without changing your hip-hinge position. Target the kettlebell to the small triangle above your knees.
5. Stand up with the kettlebell until you are fully erect, squeezing your gluteal muscles hard, pulling your kneecaps up strongly, and bracing your abdominal muscles hard. As you are standing up, keep your humerus next to your side, allowing the elbow to bend approximately 90 degrees.
6. At the top, square up your shoulders and hips.
7. Once the kettlebell starts descending, pull it down, keeping your humerus next to your side. Keep your vertical plank position until the last moment, and then get out of the way of the kettlebell. Doing this will keep the kettlebell in the small triangle between your legs during your hip hinge.
8. Complete all repetitions on one side, then switch hands.

Muscles Involved

Primary: Erector spinae (iliocostalis, longissimus, spinalis), gluteus maximus, contralateral quadratus lumborum and gluteus medius and gluteus minimus, hamstrings (semitendinosus, semimembranosus, biceps femoris), latissimus dorsi, brachioradialis, rectus abdominis, transversus abdominis, internal oblique, external oblique

Secondary: Trapezius, rhomboids, quadriceps (rectus femoris, vastus lateralis, vastus medialis, vastus intermedius), ipsilateral gluteus medius and gluteus minimus, forearms (wrist flexors, finger flexors)

Anatomic Focus

Hand spacing: Put one hand in contact with the middle of the kettlebell handle, clasping it loosely with your fingers.

Grip: Use a single overhand grip (pronate the hand).

Stance: Position your legs approximately one foot (0.3 m) behind the kettlebell and shoulder-width apart. Point your toes slightly outward.

Trajectory: Keep your elbow bent on the upswing and keep the handle of the kettlebell horizontal. The T. rex swing starts the process of getting you ready to perform the kettlebell snatch.

Range of motion: Keeping your arm straight, hike the kettlebell back between your legs, aiming for the small triangle above your knees. Pretend that you are playing American football and are snapping the ball to a friend 10 to 15 yards (9-14 m) behind you. Keep your hip-hinge position the same while hiking the kettlebell. When you have reached the end of the hike, stand up quickly. Your toes will be rooted to the ground, your kneecaps pulled up, your gluteal muscles squeezed, and your abdominal and latissimus dorsi muscles strongly engaged. As in the kettlebell deadlift, the erector spinae, abdominal muscles, and latissimus dorsi muscles will help to stabilize and straighten the spine while the gluteus maximus and hamstrings generate hip extension. Your spine should be straight and stiff throughout the movement. Antishrug your shoulders, further activating your latissimus dorsi muscles. Do not overextend your spine.

I recommend that you have the nonworking arm mimic what the working arm is doing. If you wish, when the kettlebell is at the top of the movement, you can place the nonworking hand in a guard position instead of having it out in front. Making sure that the nonworking arm is moving in this way will help avoid rotation during the exercise.

Technique note: This exercise *is not* for learning or performing the kettlebell swing. Please see the chapter on swings (chapter 3). You will learn the T. rex swing to progress from this movement to the kettlebell snatch. Because the kettlebell snatch is a vertical projection of force, the closer the kettlebell is to your body, the less energy is required to snatch it.

Switch your start hand at the beginning of each succeeding set. Doing this prevents having a favorite start hand and will help in keeping a balanced body while training.

SNATCH

SINGLE KETTLEBELL SNATCH WITH ONE HAND

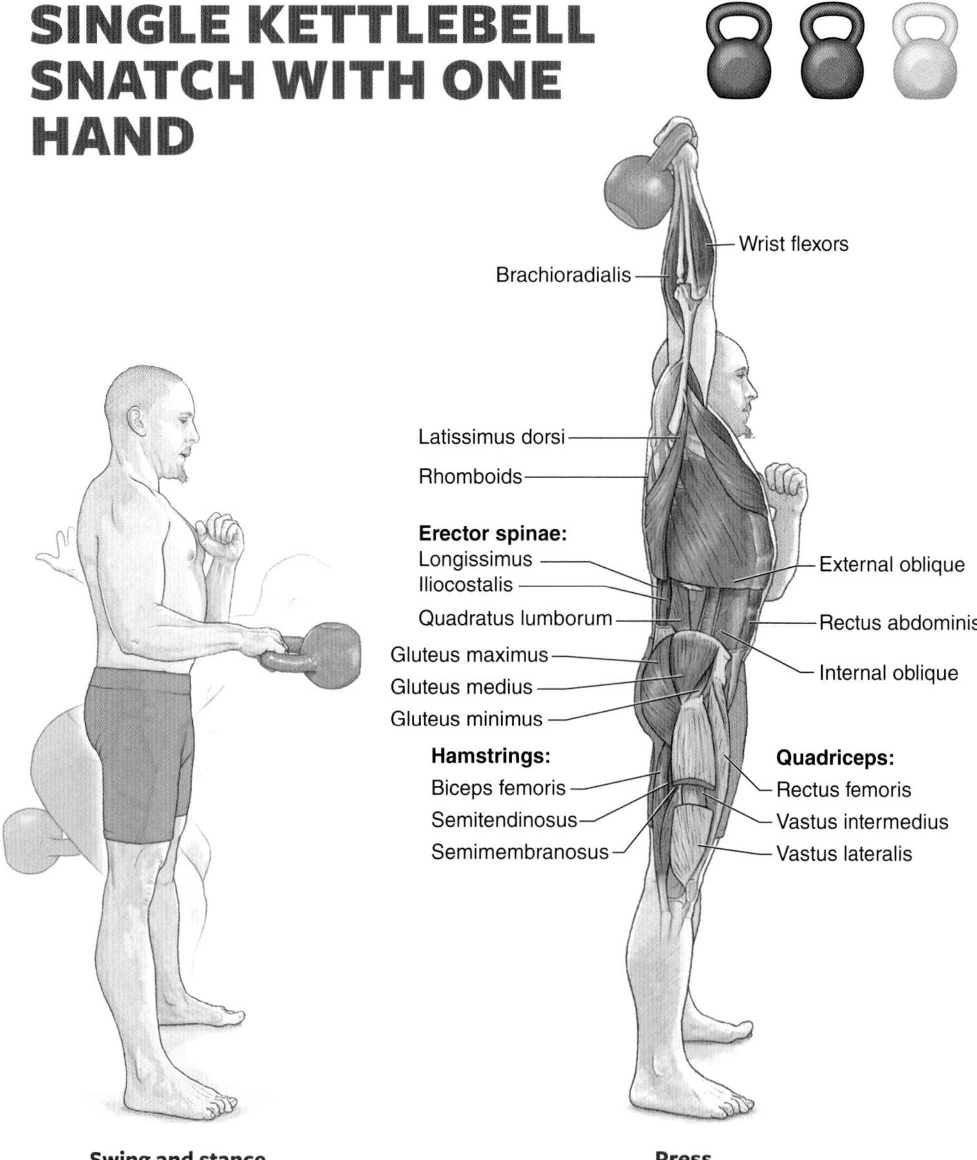

Swing and stance.

Press.

Execution

1. Place your feet about a foot (0.3 m) behind the kettlebell. Take a shoulder-width stance behind the kettlebell with your toes slightly angled outward.
2. Keeping your hips above your knees and below your shoulders, grip the kettlebell handle with one hand, placing your body into a hip-hinge position. You should feel tension in your hamstrings. Tilt the handle toward you, allowing it to become an extension of your arm.

3. Place your nonworking hand beside the kettlebell but not on it. Doing this will help keep your shoulders square at the beginning.
4. Grip the ground with your toes, tighten the latissimi dorsi, abdominal muscles, and grip, and hike pass the kettlebell back between your legs without changing your hip-hinge position. Target the kettlebell to the small triangle above your knees.
5. Stand up with the kettlebell until you are fully erect, squeezing your gluteal muscles hard, pulling your kneecaps up strongly, and bracing your abdominal muscles hard.
6. As you are standing up, perform the T. rex swing as before but allow the kettlebell to keep moving vertically. As you are doing this, the body of the kettlebell will start to flip over the hand. As this is happening, punch your hand toward the ceiling. When your elbow is straight and your shoulder is flexed, the kettlebell will be resting on the back side of your forearm and your wrist will be neutral. Your upper arm will be close to your ear.
7. At the top, be sure your shoulders and hips are squared up.
8. On the descent, pull your elbow down, allowing the kettlebell to flip over. Keep your vertical plank position until the last moment, and then get out of the way of the kettlebell. Doing this will keep the kettlebell in the small triangle above your knees during your hip hinge.
9. Complete all repetitions on one side, then switch hands.

Muscles Involved

Primary: Erector spinae (iliocostalis, longissimus, spinalis), gluteus maximus, contralateral quadratus lumborum and gluteus medius and gluteus minimus, hamstrings (semitendinosus, semimembranosus, biceps femoris), anterior deltoid, pectoralis major, latissimus dorsi, brachioradialis, rectus abdominis, transversus abdominis, internal oblique, external oblique, forearms (wrist flexors, finger flexors)

Secondary: Trapezius, rhomboids, quadriceps (rectus femoris, vastus lateralis, vastus medialis, vastus intermedius), ipsilateral gluteus medius and gluteus minimus

Anatomic Focus

Hand spacing: Put one hand in contact with the middle of the kettlebell handle, clasping it loosely with your fingers.

Grip: Use a single overhand grip (pronate the hand).

Stance: Position your legs approximately one foot (0.3 m) behind the kettlebell and shoulder-width apart. Point your toes slightly outward.

(continued)

SINGLE KETTLEBELL SNATCH WITH ONE HAND *(continued)*

Trajectory: The kettlebell snatch is a vertical projection of force, similar to the kettlebell clean. Keep the kettlebell close to your body on both the upward and downward parts of the snatch. Imagine there is a curtain two feet (0.6 m) in front of you that you don't want to touch with the kettlebell.

Range of motion: Keeping your arm straight, hike the kettlebell back between your legs, aiming for the small triangle above your knees. Pretend that you are playing American football and are snapping the ball to a friend 10 to 15 yards (9-14 m) behind you. Keep your hip-hinge position the same while hiking the kettlebell. When you have reached the end of the hike, stand up quickly. Your toes will be rooted to the ground, your kneecaps pulled up, your gluteal muscles squeezed, and your abdominal muscles and your latissimi dorsi strongly engaged. As in the kettlebell deadlift, the erector spinae, abdominal muscles, and latissimus dorsi muscles will help to stabilize and straighten the spine while the gluteus maximus and hamstrings generate hip extension. Your spine should be straight and stiff throughout the movement. Antishrug your shoulders, further activating your latissimus dorsi muscles. Do not overextend your spine.

I recommend that you have your nonworking arm mimic what the working arm is doing. If you wish, when the kettlebell is at the top of the movement, you can place the nonworking arm in a guard position instead of having it out in front. Making sure that the nonworking arm is moving in this way will help avoid rotation during the exercise. Another position that some use is to have the nonworking arm out to the side when the working arm is at the top.

Technique note: The kettlebell snatch is a vertical projection of force. The closer the kettlebell is to your body, the less energy required to snatch it. I recommend that you perform fewer snatches in the beginning, perhaps sets of three to five repetitions. Build your snatching skill slowly and increase your capability and experience, increasing your sets before your repetitions. As you move the kettlebell up, escort it into position above your head with your hand and arm. The bottom part of the swing, clean, and snatch are all indistinguishable.

Switch your start hand at the beginning of each succeeding set. Doing this prevents having a favorite start hand and will help in keeping a balanced body while training.

A few words of caution: If you are not able to obtain the proper overhead position, do not perform the kettlebell snatch.

DOUBLE KETTLEBELL SNATCH

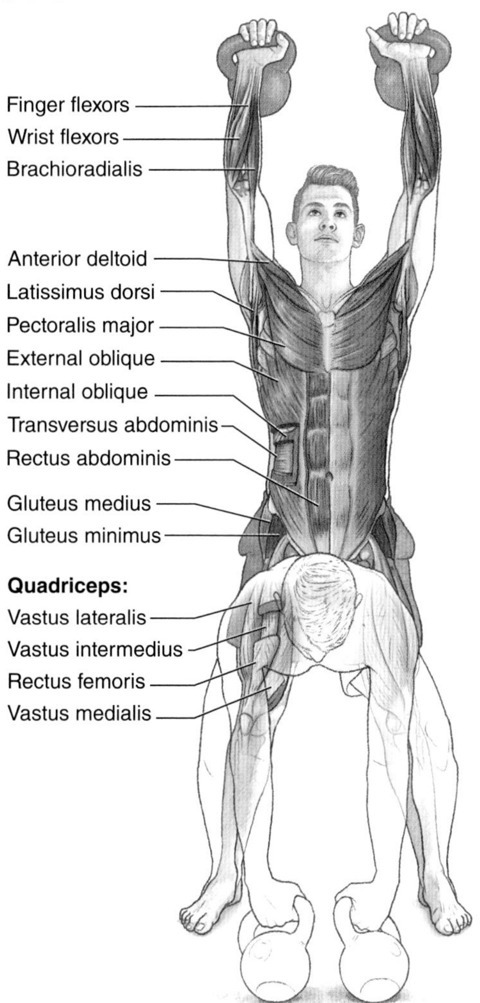

Execution

1. Place your feet about a foot (0.3 m) behind the kettlebells. Take a slightly wider than shoulder-width stance behind the kettlebells with your toes slightly pointed outward.
2. Keeping your hips above your knees and below your shoulders, grip the kettlebell handles with your hands, placing your body into a hip-hinge position. You should feel tension in your hamstrings. Tilt the handle toward you, allowing it to become an extension of your arm.
3. Grip the ground with your toes, tighten your latissimi dorsi, your abdominal muscles, and your grip, and hike pass the kettlebells back between your legs without changing your hip-hinge position. Target the kettlebells to the small triangle above your knees.

(continued)

DOUBLE KETTLEBELL SNATCH *(continued)*

4. Stand up with the kettlebells until you are fully erect, squeezing your gluteal muscles hard, pulling your kneecaps up strongly, and bracing your abdominal muscles hard.
5. As you are standing up, perform the T. rex swing as before but allow the kettlebells to keep moving vertically. As you are doing this, the body of the kettlebells will start to flip over the hand. As this is happening, punch your hands toward the ceiling. When your elbows are straight and your shoulders are flexed, the kettlebells will be resting on the back sides of your forearms and your wrists will be neutral. Your upper arms will be close to your ears.
6. At the top, make sure your shoulders and hips are squared up.
7. On the descent, pull your elbows down, allowing the kettlebells to flip over. Keep your vertical plank position until the last moment, and then get out of the way of the kettlebells. Doing this will keep the kettlebells in the small triangle above your knees during your hip hinge. **Alternative descent:** From the top position, lower the kettlebells to your chest as you do in the eccentric part of the kettlebell press. From the rack position, bring the kettlebells back between your legs as you do in the kettlebell clean. From this position, snatch the kettlebells back up and repeat. This method is better for those performing the double kettlebell snatch for the first time. In addition, if you prefer to continue performing the double snatch with this method after you are familiar with the exercise, proceed.
8. Complete all repetitions and then set the kettlebells safely on the ground in reverse order of the way you got them up.

Muscles Involved

Primary: Erector spinae (iliocostalis, longissimus, spinalis), gluteus maximus, gluteus medius, gluteus minimus, hamstrings (semitendinosus, semimembranosus, biceps femoris), anterior deltoid, pectoralis major, latissimus dorsi, brachioradialis, rectus abdominis, transversus abdominis, internal oblique, external oblique, forearms (wrist flexors, finger flexors)

Secondary: Trapezius, rhomboids, quadriceps (rectus femoris, vastus lateralis, vastus medialis, vastus intermedius)

Anatomic Focus

Hand spacing: Put each hand on its own kettlebell.

Grip: Use a double overhand grip (pronate both hands).

Stance: Position your legs approximately one foot (0.3 m) behind the kettlebells and slightly more than shoulder-width apart. Point your toes slightly outward.

Trajectory: The double kettlebell snatch is a vertical projection of force similar to the double kettlebell clean. Keep the kettlebells close to your body on both the ascent and descent parts of the snatch. Imagine there is a curtain two feet (0.6 m) in front of you that you don't want to touch with the kettlebells.

Range of motion: Whether you pull the kettlebells down to the rack position or bring them down as you would in a single kettlebell snatch, you are still doing the same part on the ascent: building a strong posterior chain. If you bring the kettlebells down in one motion, pay attention to your lower back and torso in general. Stop the set before fatigue sets in. When performing the double kettlebell snatch, use kettlebells one or two sizes lighter than your single kettlebell snatch weight.

Keeping your arms straight, hike the kettlebells back between your legs, aiming for the small triangle above your knees. Pretend that you are playing American football and are snapping the ball to a friend 10 to 15 yards (9-14 m) behind you. Keep your hip-hinge position the same while hiking the kettlebells. When you have reached the end of the hike, stand up quickly. Your toes will be rooted to the ground, your kneecaps pulled up, your gluteal muscles squeezed, and your abdominal muscles and your latissimi dorsi strongly engaged. As in the kettlebell deadlift, the erector spinae, abdominal muscles, and latissimus dorsi muscles will help to stabilize and straighten the spine while the gluteus maximus and hamstrings generate hip extension. Your spine should be straight and stiff throughout the movement. Antishrug your shoulders, further activating your latissimus dorsi muscles. Do not overextend your spine.

Technique note: Before attempting the double kettlebell snatch, you need to develop a strong double kettlebell swing, great shoulder mobility and stability, a strong mental state, and command of the single kettlebell snatch. Use kettlebells one or two sizes lighter than your single kettlebell snatch weight.

The double kettlebell snatch is a vertical projection of force. The closer the kettlebells are to your body, the less energy required to snatch it. I recommend that you perform fewer snatches in the beginning, perhaps sets of three to five repetitions. Build your snatching skill slowly and increase your capability and experience, increasing your sets before your repetitions. As you move the kettlebells upward, escorting them into position above your head with your hand and arm. The bottom part of the double swing, clean, and snatch are all indistinguishable.

SINGLE KETTLEBELL SNATCH TO ECCENTRIC PRESS

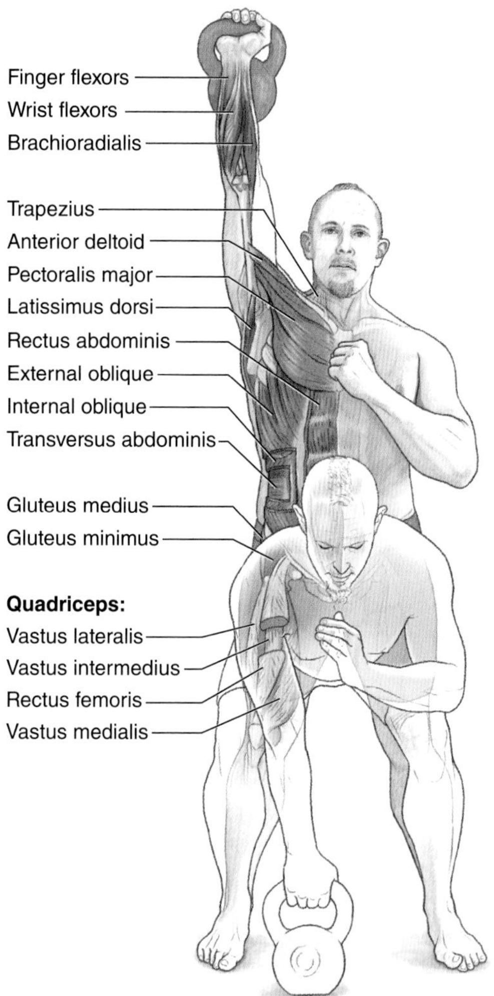

Execution

1. Place your feet about a foot (0.3 m) behind the kettlebell. Take a shoulder-width stance behind the kettlebell with your toes pointed slightly outward.
2. Keeping your hips above your knees and below your shoulders, grip the kettlebell handle with one hand, placing your body into a hip-hinge position. You should feel tension in your hamstrings. Tilt the handle toward you allowing it to become an extension of your arm.

3. Place your nonworking hand beside the kettlebell but not on it. Doing this will help keep your shoulders square at the beginning.
4. Grip the ground with your toes, tighten your latissimi dorsi, your abdominal muscles, and your grip, and hike pass the kettlebell back between your legs without changing your hip-hinge position. Target the kettlebell to the small triangle above your knees.
5. Stand up with the kettlebell until you are fully erect, squeezing your gluteal muscles hard, pulling your kneecaps up strongly, and bracing your abdominal muscles hard.
6. As you are standing up, perform the T. rex swing as before but allow the kettlebell to keep moving vertically. As you are doing this, the body of the kettlebell will start to flip over the hand. As this is happening, punch your hand toward the ceiling. When your elbow is straight and your shoulder is flexed, the kettlebell will be resting on the back side of your forearm and your wrist will be neutral. Your upper arm will be close to your ear.
7. At the top, square up your shoulders and hips.
8. On the descent, pull your elbow down slowly, bringing the kettlebell to your chest and into the rack position. You will be strongly activating your latissimus dorsi muscle at this time.
9. On the descent, keep the upper arm next to your torso while dropping the kettlebell. Keep your vertical plank position until the last moment, and then get out of the way of the kettlebell. Hike pass the kettlebell and snatch it back up to the top position.
10. Complete all repetitions on one side, then switch hands.

Muscles Involved

Primary: Erector spinae (iliocostalis, longissimus, spinalis), gluteus maximus, contralateral quadratus lumborum and gluteus medius and gluteus minimus, hamstrings (semitendinosus, semimembranosus, biceps femoris), anterior deltoid, pectoralis major, latissimus dorsi, brachioradialis, rectus abdominis, transversus abdominis, internal oblique, external oblique, forearms (wrist flexors, finger flexors)

Secondary: Trapezius, rhomboids, quadriceps (rectus femoris, vastus lateralis, vastus medialis, vastus intermedius), ipsilateral gluteus medius and gluteus minimus

Anatomic Focus

Hand spacing: Place one hand in contact with the middle of the kettlebell handle, clasping it loosely with your fingers.

Grip: Use a single overhand grip (pronate the hand).

(continued)

SINGLE KETTLEBELL SNATCH TO ECCENTRIC PRESS *(continued)*

Stance: Position your legs approximately one foot (0.3 m) behind the kettlebell and shoulder-width apart. Point your toes slightly outward.

Trajectory: The kettlebell snatch is a vertical projection of force, as is the kettlebell clean. Keep the kettlebell close to your body on the ascent of the snatch. Imagine there is a curtain two feet (0.6 m) in front of you that you don't want to touch with the kettlebell. On the descent, bring the kettlebell down to your chest in a slight diagonal pattern.

Range of motion: Keeping your arm straight, hike the kettlebell back between your legs, aiming for the small triangle above your knees. Pretend that you are playing American football and are snapping the ball to a friend 10 to 15 yards (9-14 m) behind you. Keep your hip-hinge position the same while hiking the kettlebell. When you have reached the end of the hike, stand up quickly and snatch the kettlebell overhead. Your toes will be rooted to the ground, your kneecaps pulled up, your gluteal muscles squeezed, and your abdominal and latissimus dorsi muscles strongly engaged.

Technique note: The general purpose of this movement is to strengthen your press by increasing the eccentric strength. Most people can generally snatch more weight than they can press, up to a certain weight. Slowly lower the kettlebell to your chest, accentuating the eccentric aspect, to further activate the latissimi dorsi.

As in the kettlebell deadlift, the erector spinae, abdominal muscles, and latissimus dorsi muscles will help to stabilize and straighten the spine while the gluteus maximus and hamstrings generate hip extension. Your spine should be straight and stiff throughout the movement. Antishrug your shoulders, further activating your latissimus dorsi muscles. Do not overextend your spine.

I recommend that you have your nonworking arm mimic what the working arm is doing. If you wish, when the kettlebell is at the top of the movement, you can place the nonworking arm in a guard position instead of having it out in front. Making sure that the nonworking arm is moving in this way will help avoid rotation during the exercise.

Be proficient in your single kettlebell snatch technique before performing this movement. The kettlebell snatch is a vertical projection of force; the closer the kettlebell is to your body the less energy is required to snatch it. As the kettlebell is moving upward, your hand and arm are escorting it into position above your head. The bottom part of the swing, clean, and snatch are all indistinguishable.

Switch your start hand at the beginning of each succeeding set. Doing this prevents having a favorite start hand and will help in keeping a balanced body while training.

A few words of caution: If you are not able to obtain the proper overhead position, do not perform the kettlebell snatch.

VARIATION

Double Kettlebell Snatch to Eccentric Press

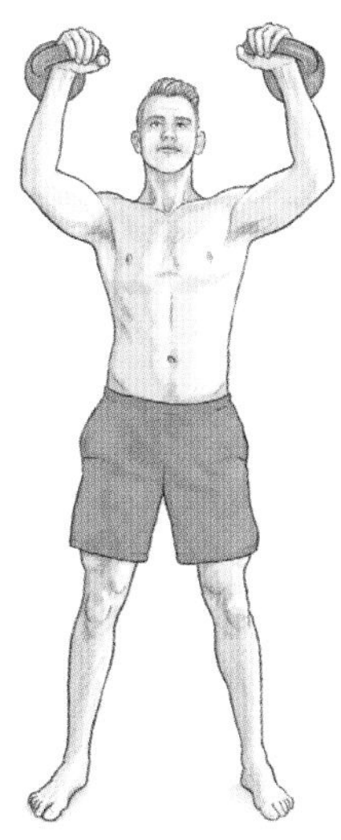

This variation is performed as the single version but with double kettlebells. Slowly lower the kettlebells to your chest from the top, accentuating the eccentric aspect, to further activate the latissimus dorsi muscles. Keep your abdominal and gluteal muscles contracted strongly throughout the eccentric phase, in addition to the double kettlebell snatch to get them to the top.

8 ROW AND PULL-UP

Core. Lift up. Lats and traps. Grip. These are words that adequately describe the row and pull-up movements, which are strong movements for the body and grind exercises that will forge a strong and wide back and develop an iron core. Your grip will become hardened, tougher, and a lot stronger.

Rows and pull-ups are not the most sought-after exercises, but they are some of the most useful. One of the main muscles that they train and strengthen is the latissimus dorsi. Your latissimi dorsi are among very few muscles that attach to an extremity and the spinal region. Therefore, training with rows and pull-ups is very important to having a well-rounded training program and staying injury free.

The latissimus dorsi not only pulls the humerus back, but it also rotates it medially. When the hands are fixed and not moving (a closed kinetic chain movement), the latissimus dorsi muscles pull the torso up and forward. Another strong action of the latissimi dorsi is stabilizing the lower spine with maximal contraction. When you antishrug prior to your exercises, I can almost guarantee that you won't round your lower back. This muscle also depresses your scapula, which further packs your shoulders. These two important qualities of your latissimus dorsi muscles will help to decrease injuries and increase your performance, both in the gym and outside of it.

One of my favorite exercises is the renegade row. There is so much packed into this one movement that you gain a lot from performing it. You combine both the static—the push-up plank—with the dynamic—the row. Both are fantastic strength builders on their own, but when combined, their total benefit exceeds their summation. Build up with this exercise; it is a grind.

The standing rows are also awesome exercises. Performing them with one hand or both hands will further strengthen your back muscles in addition to your torso muscles.

I have shared with many people that I am personally pull-up challenged. They are difficult for me. I learned at an early age that I was much better at performing push-ups than pull-ups. However, even with my disdain for performing pull-ups, I still did them. One of my powerlifting friends told me early on in my powerlifting career that getting stronger and better at doing pull-ups would not only help my deadlift but also make my bench press stronger. He was right. In this chapter, the pull-up with a kettlebell is discussed. You can also do it as a body-weight exercise without the kettlebell. Both have their merits and good training qualities.

Pull-ups are hard to perform and hard mentally. They are definitely grind movements. They will also give you a strong grip, as you are hanging and eventually pulling with your hands as they are the only thing attaching you to the bar. Practice the setup before you actually do a pull-up. You will be happy that you did. The pull-up is considered a whole-body exercise, from your toes to your hands. Obviously, your latissimus dorsi muscles are doing a lot of the work, but other muscles are also helping them or stabilizing the rest of your body to assist in the movement.

Rows and pull-ups are essential in a well-rounded training program. Not only do they make you stronger overall, but they will help your posture and counteract the anterior forces on our bodies in today's electronic world. Even if you are like me and initially disdained doing pull-ups, you will learn to love them and the rows along with their eventual benefits.

EXERCISE FOCUS FOR KETTLEBELL ROWS AND PULL-UPS

The kettlebell rows and pull-up provide different ways to help you learn proper technique and develop strength. Here, we take a closer look at the focus of the kettlebell rows and pull-up.

The exercise focus for this chapter is not only about pulling movements but also the continuation of developing a very strong core. The renegade row is one of my all-time favorite exercises, because it accomplishes many different things while I am performing the movement. In addition to developing a strong core, which will help you in many sports including ice hockey, water skiing, track and field, gymnastics, and soccer (figure 8.1), you will learn how to develop diagonal strength, especially when lifting one hand off the floor during the exercise.

Keeping both feet on the ground while doing the renegade row will also help develop antirotation strength. Creating and holding a horizontal plank position while you are lifting is extremely important as you are combining a static with a dynamic exercise. In addition, you will also strengthen your antirotation durability, especially when you go from four points of contact to three and back and learn proper tension-developing skills that will be used with other strength movements.

The standing single kettlebell row not only helps you to develop your back muscles, but it also helps you to keep your shoulders and hips squared up throughout the movement even though you are using an asymmetrical load. Since you are using an asymmetrical load during this row, you will be recruiting additional contralateral muscles such as the contralateral erector spinae, gluteus medius and gluteus minimus, and the quadratus lumborum. An added benefit of this exercise is that you will be creating and holding a static hip-hinge position while you are lifting. It is a static exercise combined with a dynamic exercise.

FIGURE 8.1 By developing diagonal strength, a soccer player will be able to kick the ball harder and farther.

For the standing double kettlebell row, you will be using a heavier weight, thereby requiring a stronger isometric contraction to be able to maintain the hinge position while doing the rows.

The final exercise is a pull-up with a kettlebell. Before performing this exercise, make sure you can perform body-weight pull-ups with good technique for 8 to 10 repetitions. Performing pull-ups with a kettlebell develops your back muscles, which are used for proper posture and for many sports such as powerlifting, weightlifting, rock climbing, archery, and tennis and for increasing your overall pulling strength. You will also create a better and stronger grip to help you perform better in your daily activities.

RENEGADE ROW

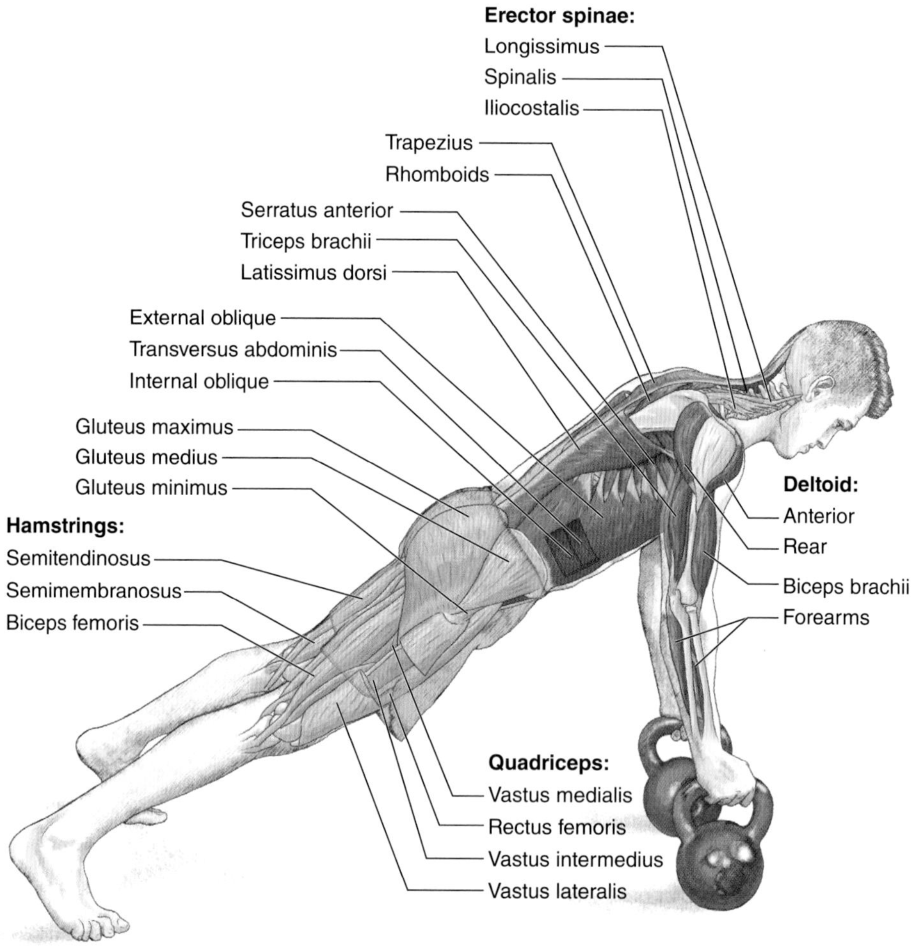

Erector spinae:
Longissimus
Spinalis
Iliocostalis
Trapezius
Rhomboids
Serratus anterior
Triceps brachii
Latissimus dorsi
External oblique
Transversus abdominis
Internal oblique
Gluteus maximus
Gluteus medius
Gluteus minimus
Hamstrings:
Semitendinosus
Semimembranosus
Biceps femoris
Deltoid:
Anterior
Rear
Biceps brachii
Forearms
Quadriceps:
Vastus medialis
Rectus femoris
Vastus intermedius
Vastus lateralis

Start position.

NOTE: For the stationary arm, use a 20-kilogram (44 lb) kettlebell or heavier due to the larger base. A smaller kettlebell with a smaller base is more likely to tip over. If you are not able to do a renegade row with a heavier kettlebell, use the bigger kettlebell for the nonlifting arm and a smaller one for the lifting arm. Perform all the repetitions with the smaller kettlebell and then switch the kettlebells and perform repetitions with the other arm.

Row.

Execution

1. Place two kettlebells on the ground with the handles parallel to each other. Assume a push-up plank position with your hands on the handles.
2. Your shoulders should be over your hands. Place your feet shoulder-width or slightly wider apart. Push back on your feet so your ankles are slightly dorsiflexed.
3. Tuck your pelvis, tighten and brace your abdominal muscles, and squeeze your gluteal muscles strongly.
4. While you are actively pushing the kettlebells into the ground, lift the left kettlebell off the ground by flexing your elbow and rowing it to your waist. Lower it back to the ground and repeat on the other side, alternating arms with each repetition.
5. Imagine there is a flag attached to your waist behind you. The goal is to not wave the flag while performing the renegade row. This means don't wiggle your hips side to side or up and down during the exercise. There should be a straight line from your head to your ankles throughout the movement.

Muscles Involved

Primary: Latissimi dorsi, trapezius, rhomboids, serratus anterior, anterior and rear deltoid, triceps brachii, biceps brachii, forearms (wrist flexors, finger flexors), rectus abdominis, external oblique, internal oblique, transversus abdominis

Secondary: Erector spinae (iliocostalis, longissimus, spinalis), gluteus maximus, gluteus medius, gluteus minimus, hamstrings (semitendinosus, semimembranosus, biceps femoris), quadriceps (rectus femoris, vastus lateralis, vastus medialis, vastus intermedius)

Anatomic Focus

Hand spacing: Put each hand in contact with the middle of the kettlebell handle, gripping it strongly.

Grip: Use a double overhand grip. Either have your hands parallel to each other or have them in an upside-down "V" position.

Stance: Position your feet shoulder-width or wider apart. Dorsiflex your ankles with your kneecaps pulled up and your gluteal muscles squeezed.

Trajectory: Move the kettlebell in a slight diagonal toward the hip.

Range of motion: The renegade row combines two exercises: the kettlebell row and the push-up plank. This movement is great for developing torso strength and stability while going from four points of contact with the ground to three points and back to four. It also helps to develop and strengthen your antirotation ability while you are lifting the kettlebell.

(continued)

RENEGADE ROW *(continued)*

Keeping your pelvis tucked (or your belly button pulled up toward your chin), tightening and bracing your abdominals, and squeezing your gluteal muscles strongly are very important to maintain the torso rigidity required to hold the body still while performing the row. Your spine should be straight and stiff throughout the movement. Antishrug your shoulders, further activating your latissimus dorsi muscles. Do not extend your spine. My recommendation is to strongly drive the kettlebell into the ground with your nonworking arm while digging your toes in as well.

This is an intense exercise, so keep your sets and repetitions down to two or three sets of three to five repetitions per side. You will be glad you did. Quality is more important than quantity. A simple regression for this exercise is to perform a push-up plank with no weight. Alternate by bringing one hand off the ground and eventually the elbow up, as you would in the renegade row.

STANDING SINGLE KETTLEBELL ROW

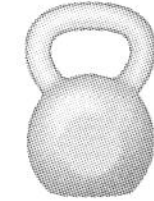

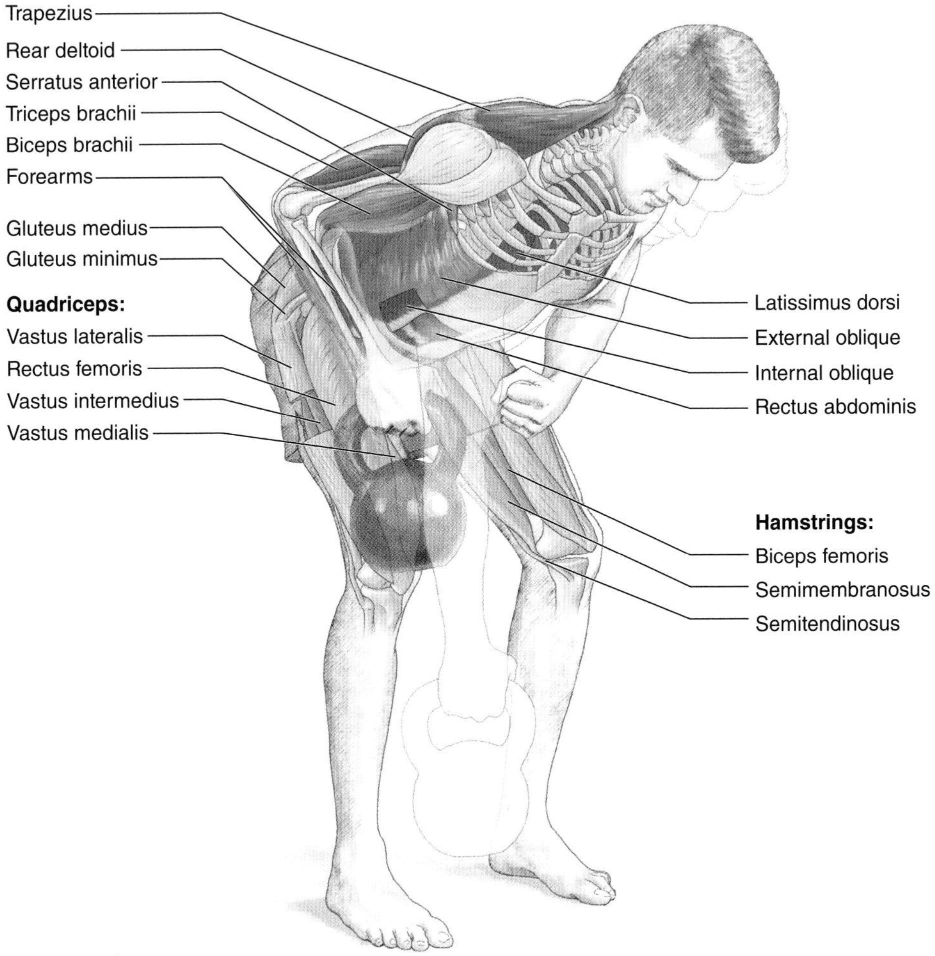

Execution

1. Take a shoulder-width stance over the kettlebell with your toes pointed slightly outward. Position the handle just in front of the legs and perpendicular to the shoulders.
2. Keeping the hips above the knees and below the shoulders, grip the kettlebell handle with one hand, placing your body into a hip-hinge position, keeping a neutral spine. Keep the nonworking hand to your side. You should feel tension in your posterior chain, namely the hamstrings and gluteal muscles.
3. Grip the ground with your toes; tighten your latissimi dorsi, your abdominal muscles, and your grip; and lift the kettlebell off the ground by

(continued)

STANDING SINGLE KETTLEBELL ROW *(continued)*

flexing your elbow and rowing it to your waist. Lower the kettlebell and repeat. Keep your hips and shoulders squared up throughout this movement. Finish all repetitions on that side then set the kettlebell on the ground. Repeat on the other side.

4. Imagine there is a flag attached to your waist behind you. The goal is to not wave the flag while performing the single kettlebell row. This means don't wiggle your hips side to side or up and down more during the exercise. Keep the angle of your torso with the ground the same for each repetition.

Muscles Involved

Primary: Latissimus dorsi, trapezius, rhomboids, serratus anterior, rear deltoid, triceps brachii, biceps brachii, forearms (wrist flexors, finger flexors), rectus abdominis, external oblique, internal oblique, transversus abdominis

Secondary: Erector spinae (iliocostalis, longissimus, spinalis), gluteus maximus, gluteus medius, gluteus minimus, hamstrings (semitendinosus, semimembranosus, biceps femoris), quadriceps (rectus femoris, vastus lateralis, vastus medialis, vastus intermedius), serratus anterior

Anatomic Focus

Hand spacing: Place one hand on the handle of the kettlebell and the other hand on your side.

Grip: Use a single overhand grip.

Stance: The legs will be approximately shoulder-width apart with the kettlebell between the legs. Point your toes slightly outward.

Trajectory: Move the kettlebell in a slight diagonal toward the hip.

Range of motion: The standing single kettlebell row combines several movements: It is a grip exercise, a fantastic torso strengthener, an antirotation training movement, and finally, a back developer. Assume the hip-hinge position from the kettlebell deadlift and row the kettlebell to your waist and back down. Repeat for the number of repetitions that you are doing and then switch hands and repeat. Keep a neutral spine during the movement with a strong brace of your abdominals and torso musculature. Keep your spine straight and stiff during the static hip hinge. The erector spinae, abdominals, and latissimus dorsi muscles will help to stabilize the spine while the lower-body musculature maintains the exercise position. Antishrug your shoulders, further activating your latissimus dorsi muscles.

This is an intense exercise, so keep your numbers of sets and repetitions low, two or three sets of five to eight repetitions per side. You will be glad you did. Aim for quality over quantity.

VARIATION

Standing Double Kettlebell Row

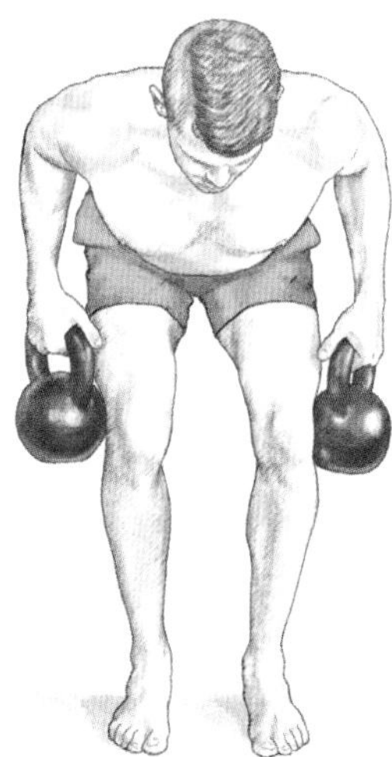

This variation is the same as the standing single kettlebell row but is performed with double kettlebells. Put each hand on a kettlebell. Simultaneously lift the kettlebells off the ground by rowing your elbows toward your waist. Lower and repeat. Work hard to keep your torso rigid and maintain the hip-hinge position throughout the exercise. If you keep the kettlebells between your legs, widen your feet. If they are outside your legs, narrow your stance slightly.

PULL-UP WITH A KETTLEBELL

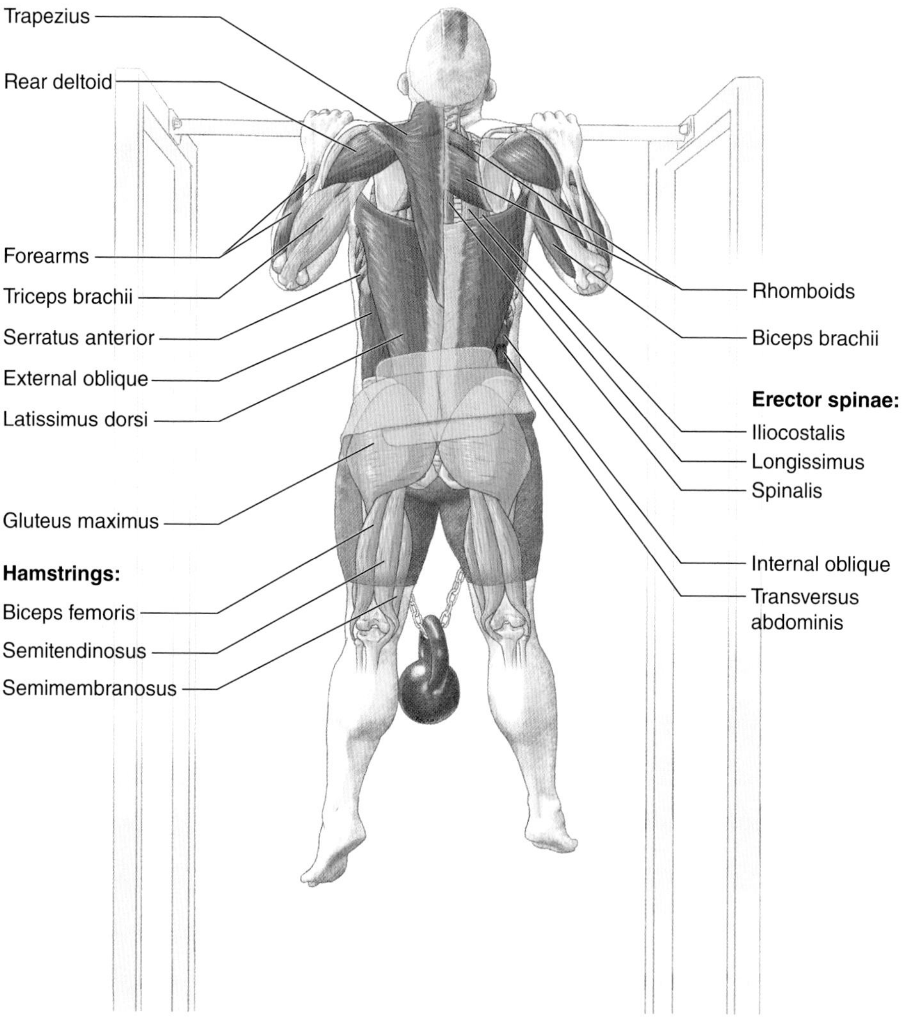

Trapezius
Rear deltoid
Forearms
Triceps brachii
Serratus anterior
External oblique
Latissimus dorsi
Gluteus maximus
Hamstrings:
Biceps femoris
Semitendinosus
Semimembranosus
Rhomboids
Biceps brachii
Erector spinae:
Iliocostalis
Longissimus
Spinalis
Internal oblique
Transversus abdominis

Execution

1. Use a dip belt to attach a kettlebell to your waist. Keep the kettlebell as high as possible to prevent it from swinging during the pull-up. (Alternatively, instead of using a dip belt to attach a kettlebell to yourself, you can hook a kettlebell with your foot and then perform a pull-up.)

2. Stand under a pull-up bar and grab the bar with a slightly wider than shoulder-width grip. An overhand thumbless grip is preferred.
3. Brace your abdominals and get into the hollow position. Squeeze your gluteal muscles strongly and tighten your lower body, getting ready for the start.
4. Assuming a dead hang position with your elbows locked out, pull yourself up. Drive your elbows back and down. Squeeze and depress your shoulder blades together. Touch your neck or upper chest to the bar.
5. Lower yourself under control back to the dead hang position.

Muscles Involved

Primary: Latissimus dorsi, trapezius, rhomboids, rear deltoid, biceps brachii, forearms (wrist flexors, finger flexors), rectus abdominis, external oblique, internal oblique, transversus abdominis

Secondary: Erector spinae (iliocostalis, longissimus, spinalis), serratus anterior, triceps brachii, gluteus maximus, hamstrings (semitendinosus, semimembranosus, biceps femoris), quadriceps (rectus femoris, vastus lateralis, vastus medialis, vastus intermedius)

Anatomic Focus

Hand spacing: Slightly wider than shoulder-width grip. An overhand thumbless grip is preferred.

Grip: Use a double overhand grip. Other options are a double supinated grip or a parallel grip.

Trajectory: Move the kettlebell, along with your body, in a vertical direction up.

Range of motion: The pull-up is one of the best exercises to strengthen your back muscles and increase hypertrophy. It also benefits other exercises that use the latissimus dorsi as part of the lift—namely the kettlebell swing and press—and the power lifts: the back squat, the bench press, and the deadlift.

During the pull-up, you will be using your whole body. Keeping the hollow position, tightening and bracing your abdominals and squeezing your gluteal muscles strongly are all very important to maintain the torso rigidity required to hold the body still while performing the pull-up. You will also develop a stronger grip performing pull-ups. Antishrug your shoulders, further activating your latissimus dorsi muscles. Quality over quantity. A simple regression for this exercise is to perform a pull-up with no weight.

9

CARRY

Simple. Hard. Tough. Core and grip on fire. Attention to detail. Calorie burner. These words adequately describe the group of movements called carries. In this chapter we will be talking about four types of carries: farmer's carry, rack carry, overhead carry, and bottom-up carry.

Performing carries other than the farmer's carries requires you to become familiar with earlier chapters of this book: the clean and press chapter (chapter 4) and the snatch chapter (chapter 7). Do farmer's carries for some time before progressing to the other carries with those same parts (clean, press, and snatch).

The carries are simple to perform: You just walk with the weight. However, in doing so, you will be breathing hard, and your heart rate will go up. Who knew that this exercise would enhance your work capacity and improve your cardiovascular endurance? Regardless of how you are carrying the weight, you are still walking under load. It's one thing to lift the weight with your feet stationary, but another thing to move your feet continuously under load. Keep your breathing under control, even when breathing heavy. Carries are simple and hard at the same time.

Mental toughness is a quality that will improve when you are doing carries. If you are performing farmer's carries while you are going for a distance and your grip feels like it is ready to give out, crush the handle even more. That is being mentally tough. When your abdominal muscles are on fire by keeping your ribs from flaring out and your lower back from extending while performing the overhead carry, you cramp your abdominal muscles harder and finish the time. This exercise will forge an iron will.

As in the get-up (chapter 5) and clean and press (chapter 4), your grip will definitely get a workout with the carries. The interesting thing about the grip is that when you are crushing something, it sends a signal up the body's pipeline telling it something big is coming. This further tightens your elbow and shoulder musculature, followed by the spinal stabilizers, including the abdominal and the latissimus dorsi muscles. That is why your core and grip will be on fire during this exercise.

Keep the numbers of sets and repetitions low for carries; two or three sets are plenty. If you are doing carries as part of a warm-up, keep the weight low or medium. If carries are part of your training session, go heavier. If they are for the end, medium to heavy weight is best, based on what you did earlier in the session. For distance, start at 20 to 30 yards (18-27 m) and build up. If you are going for time, go for 30 seconds and slowly increase the time. As you are walking with a load, make sure the area that you are walking in is clear and nothing is in your way. These are safe exercises if done properly.

Attention to detail is important when performing carries. This is a dynamic exercise, and you are moving, therefore always know what is front of you. Keep your eyes on the horizon. Be constantly checking on your abdominal and gluteal muscles. Make sure they are firing all the time while walking. Keep your posture and the vertical plank position while moving forward. If you are unable to do this for the entire distance, set the kettlebells down at that spot, take a few deep breaths, and then continue. Quality over quantity.

An important positive side effect of performing carries and other kettlebell exercises is developing excess postexercise oxygen consumption. This is an increased rate of oxygen intake that occurs immediately following exercise. You burn calories during exercise and even for a time after exercise. While you are walking with a load, your heart rate increases, and you start breathing harder. You will burn calories while making your muscular system stronger as you are recruiting almost every muscle in the body. In the end, you will feel better and look better.

Carries are not complex, but they are hard. They burn calories and make you a better person overall. Start out with farmer's carries and perform them for some time. Try double kettlebells and then a single kettlebell. Switch sides when doing the singles. Work your way up to rack, overhead, and bottom-up carries and enjoy the hard work and the great results that will come with them.

EXERCISE FOCUS FOR KETTLEBELL CARRIES

Each of the kettlebell carries provides different ways to help you develop strength. Here, we take a closer look at the focus of the kettlebell carries.

Walking is one of the most primal movements that humans perform daily. When you combine this movement with weight, it not only enables you to improve your nonweighted walking, but it also helps your overall posture while standing. The double kettlebell farmer's carry benefits you by helping to develop a strong core, improving your grip strength and mental toughness and improving your cardiovascular system and your work capacity, which will help you to handle higher workloads for a longer time. Sports that benefit from this exercise include those where strength endurance is a premium, such as cross-country skiing, football, martial arts, and cycling (figure 9.1).

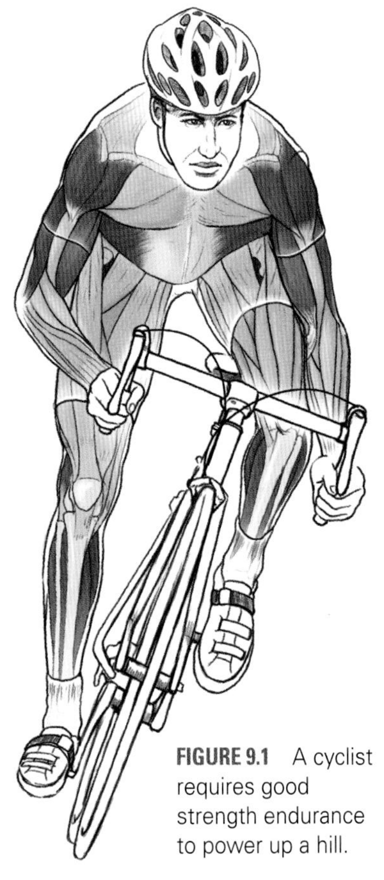

FIGURE 9.1 A cyclist requires good strength endurance to power up a hill.

When you are walking, you often have only one foot on the ground. Walking with weight helps you improve your balance and overall motor function. A positive side effect of walking with kettlebells is that you will strengthen your rotator cuff as it centralizes the humerus, allowing the stabilizer muscles to work hard. This is in addition to learning proper tension-developing skills that will be used with other strength movements. The single kettlebell farmer's carry is an advanced progression of the double kettlebell farmer's carry in that you load your body asymmetrically while maintaining proper posture and walking. Keeping your shoulders and hips squared up throughout this movement is paramount to deriving the benefits of it.

The next three types of carries—rack carry, overhead carry, and bottom-up carry—are advanced progressions of the farmer's carry in that the overall center of mass is elevated higher than the farmer's carry. Whether performing the kettlebell rack carry with double or single kettlebells, maintaining proper body posture is extremely important to help prevent injury and improve your overall performance in your sport. The kettlebell overhead carry is a more advanced progression; the kettlebell is above your head, and it is challenging to keep it overhead for the exercise. Having the requisite mobility to hold a kettlebell above your head is very important prior to performing this exercise, whether with double or single kettlebells.

As with any single kettlebell movements, asymmetrically loading yourself causes the contralateral side to work harder. An additional advantage to the kettlebell bottom-up carry is that you will develop a powerful grip and rugged mental toughness.

DOUBLE KETTLEBELL FARMER'S CARRY

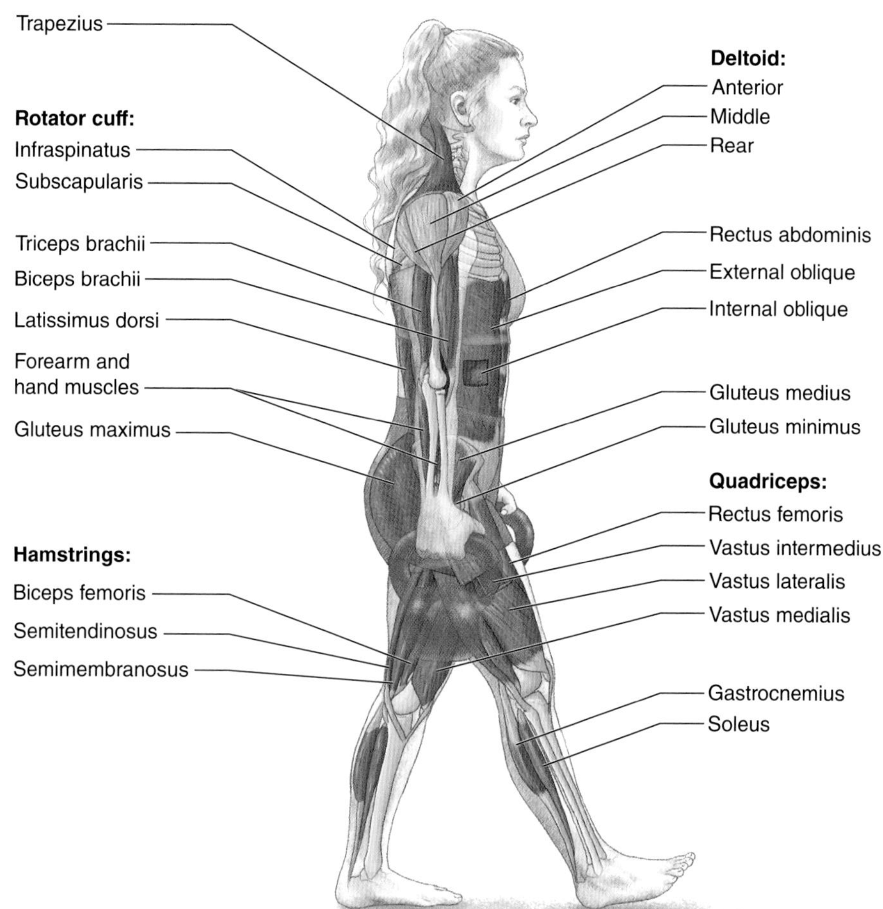

Execution

1. Place two kettlebells on the ground with the handles parallel to each other and far enough apart to put your feet between them.
2. Place your body into a hip-hinge position, similar to what you do in the kettlebell deadlift. Reach down and grip the kettlebells. Deadlift the kettlebells and stand straight. Assume the vertical plank position.
3. Walk the prescribed distance. Keep your spine as neutral as possible and hold the vertical plank position while walking. In addition, your shoulders, back, and abdominal muscles should be tight. Your stride length will be shorter than normal while performing the kettlebell farmer's carry.

4. When you have reached the prescribed distance, set the kettlebells on the ground in reverse of the motion to stand up with them.
5. Rest and repeat the sets until done.

Muscles Involved

Primary: Latissimus dorsi, trapezius, rhomboids, triceps brachii, biceps brachii, forearm and hand muscles (wrist flexors, finger flexors), rectus abdominis, external and internal obliques, transversus abdominis, gluteus maximus, gluteus medius, gluteus minimus, hamstrings (semitendinosus, semimembranosus, biceps femoris), quadriceps (rectus femoris, vastus lateralis, vastus medialis, vastus intermedius), gastrocnemius, soleus

Secondary: Erector spinae (iliocostalis, longissimus, spinalis), deltoid (anterior, middle, and rear), rotator cuff (infraspinatus, supraspinatus, teres minor, subscapularis)

Anatomic Focus

Hand spacing: Each hand is in contact with the middle of the kettlebell handle, gripping it strongly.

Grip: Use a double neutral grip and be sure your hands are parallel to each other.

Stance: Place your feet in a hip width stance with the toes pointed slightly outward with the kettlebells outside the ankles.

Trajectory: Initially, move the kettlebells in a path that is straight up and down and close to the body to get into position. Move them in a horizontal path while walking.

Range of motion: Keeping your arms extended and your elbows stiff, stand up with the kettlebells. Keep your spine straight and stiff throughout the movement. The erector spinae, abdominal muscles, and latissimus dorsi muscles will help to stabilize and straighten the spine, and the gluteus maximus and hamstrings generate hip extension. Antishrug your shoulders, further activating your latissimus dorsi muscles. Cramp your gluteal muscles to stand straight up. Do not overextend your spine. At this point, walk forward for the prescribed distance, then set the kettlebells down and rest.

Exercise note: The farmer's carry allows a lot of variability in sets, repetitions, distance, and time. You can perform it for a specific time walking (e.g., three sets of 35 seconds), distance walked (e.g., three sets of 40 yards [37 m]), and as part of your training session or as a warm-up or a finisher. A recommended weight to start with is one third to one half of your body weight during the farmer's carry. Aim for quality over quantity.

(continued)

DOUBLE KETTLEBELL FARMER'S CARRY *(continued)*

VARIATION

Single Kettlebell Farmer's Carry (Suitcase Carry)

The suitcase carry is a unilateral version of the double kettlebell farmer's carry. It is performed the same as the double kettlebell farmer's carry is but with one kettlebell. I recommend you carry the kettlebell on one side a certain time or distance. Rest, switch hands, then carry it back. There is some variability with this as well.

Just as in the one-hand swing, the contralateral musculature, namely the quadratus lumborum, gluteus medius and gluteus minimus, are working harder during the suitcase carry to keep the vertical plank position and your posture upright during the walk.

DOUBLE KETTLEBELL RACK CARRY

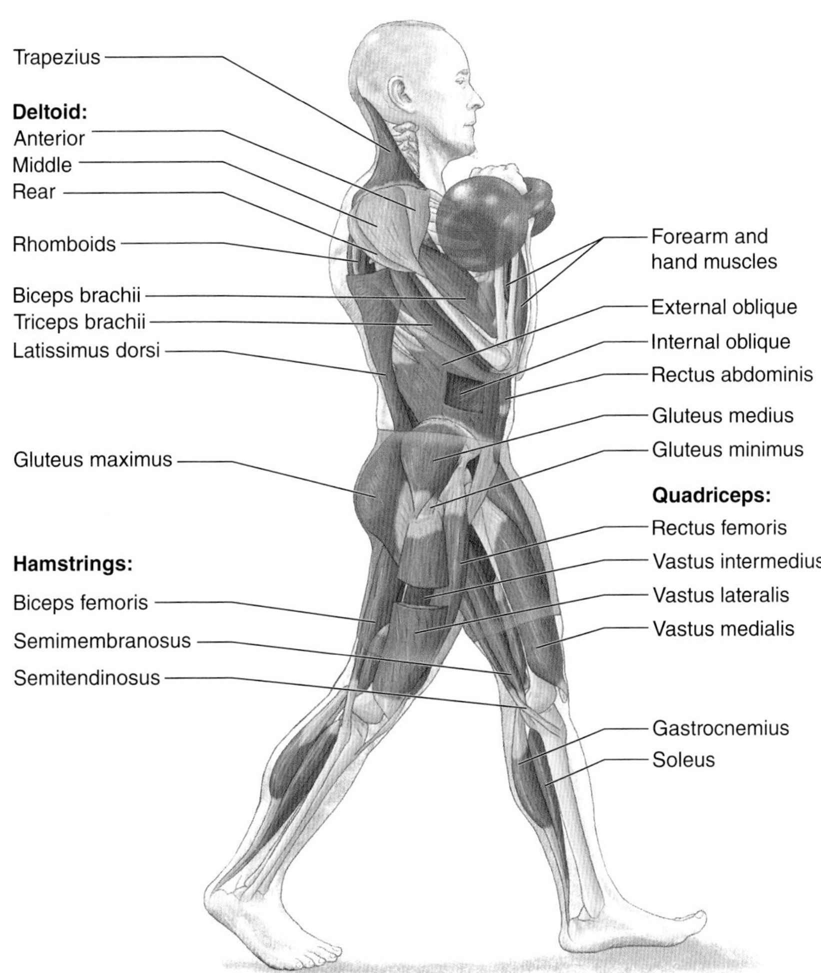

Execution

1. Place two kettlebells on the ground with the handles in line with each other, or you may make a *V* with the handles. Place your feet about a foot (0.3 m) behind the kettlebells. Take a slightly wider than shoulder-width stance behind the kettlebells with your toes pointed slightly outward.
2. Perform a double kettlebell clean. Hold the kettlebells in the rack position and move your stance in to normal walking width.

(continued)

DOUBLE KETTLEBELL RACK CARRY *(continued)*

3. Walk the prescribed distance. Keep your spine as neutral as possible and maintain the vertical plank position while walking. In addition, your shoulders, back, and abdominal muscles should be tight. Your stride length will be shorter than normal while performing the kettlebell rack carry.
4. When you have reached the prescribed distance, set the kettlebells on the ground in reverse of the motion to stand up with them.
5. Rest and repeat until done.

Muscles Involved

Primary: Latissimus dorsi, trapezius, rhomboids, triceps brachii, biceps brachii, forearm and hand muscles (wrist flexors, finger flexors), rectus abdominis, external and internal obliques, transversus abdominis, gluteus maximus, gluteus medius, gluteus minimus, hamstrings (semitendinosus, semimembranosus, biceps femoris), quadriceps (rectus femoris, vastus lateralis, vastus medialis, vastus intermedius), gastrocnemius, soleus

Secondary: Erector spinae (iliocostalis, longissimus, spinalis), deltoid (anterior, middle, and rear)

Anatomic Focus

Hand spacing: Each hand is in contact with the middle of the kettlebell handle, clasping it loosely with your fingers.

Grip: Use a double overhand grip—a hand will be on each kettlebell. Position the kettlebell handles in line with each other or make a *V* with the handles.

Stance: Position your legs approximately one foot (0.3 m) behind the kettlebell and slightly more than shoulder-width apart. Point your toes slightly outward. Once the kettlebells are in the rack position, move your feet in to normal walking width.

Trajectory: Initially, the kettlebell path should be straight up and down and close to your body to get into position. They will travel a horizontal path as you are walking.

Range of motion: While holding the kettlebells in the rack position, keep your spine straight and stiff. The erector spinae, abdominal muscles, and latissimus dorsi muscles will help to stabilize and straighten the spine, and the gluteus maximus and hamstrings will generate hip extension while you perform a kettlebell clean and then walk with them. Antishrug your shoulders, further activating your latissimus dorsi muscles. Do not overextend your spine. At this point, walk forward for the prescribed distance, then set the kettlebells down and rest.

Holding the kettlebells in the rack position raises your overall center of mass (which is based on your body weight plus the weight of the kettlebells). Doing this will place more emphasis on keeping your spine neutral and firing your abdominal and gluteal muscles strongly throughout the rack carry.

Exercise note: The rack carry allows a lot of variability in sets, repetitions, distance, and time. You can perform it for length of time walking (e.g., three sets of 35 seconds), distance walked (e.g., three sets of 40 yards [37 m]), and as part of your training session or as a warm-up or a finisher. A recommended weight to start with is one fourth to one third of your body weight during the rack carry. Aim for quality over quantity.

VARIATION

Single Kettlebell Rack Carry

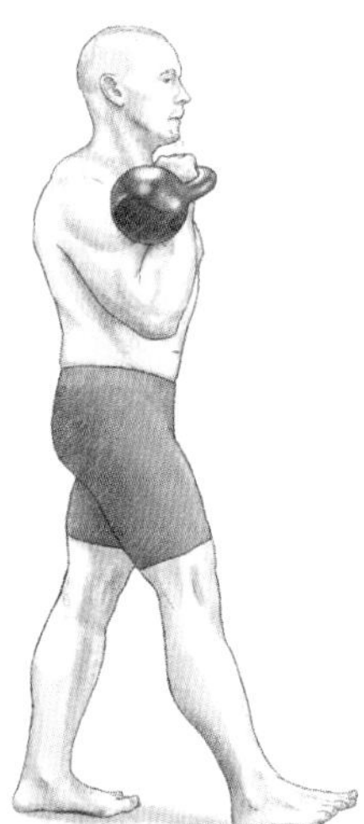

The single kettlebell rack carry is a unilateral version of the double kettlebell rack carry. It is performed the same but with one kettlebell. I recommend that you carry the kettlebell on one side a certain time or distance, rest, switch hands, then carry it back. As mentioned in previous exercises, there is some variability with this.

Similar to the one-hand swing, the contralateral musculature, namely the quadratus lumborum, gluteus medius, and gluteus minimus, are working harder during the single kettlebell rack carry to keep the vertical plank position and your posture upright during the walk.

SINGLE KETTLEBELL OVERHEAD CARRY

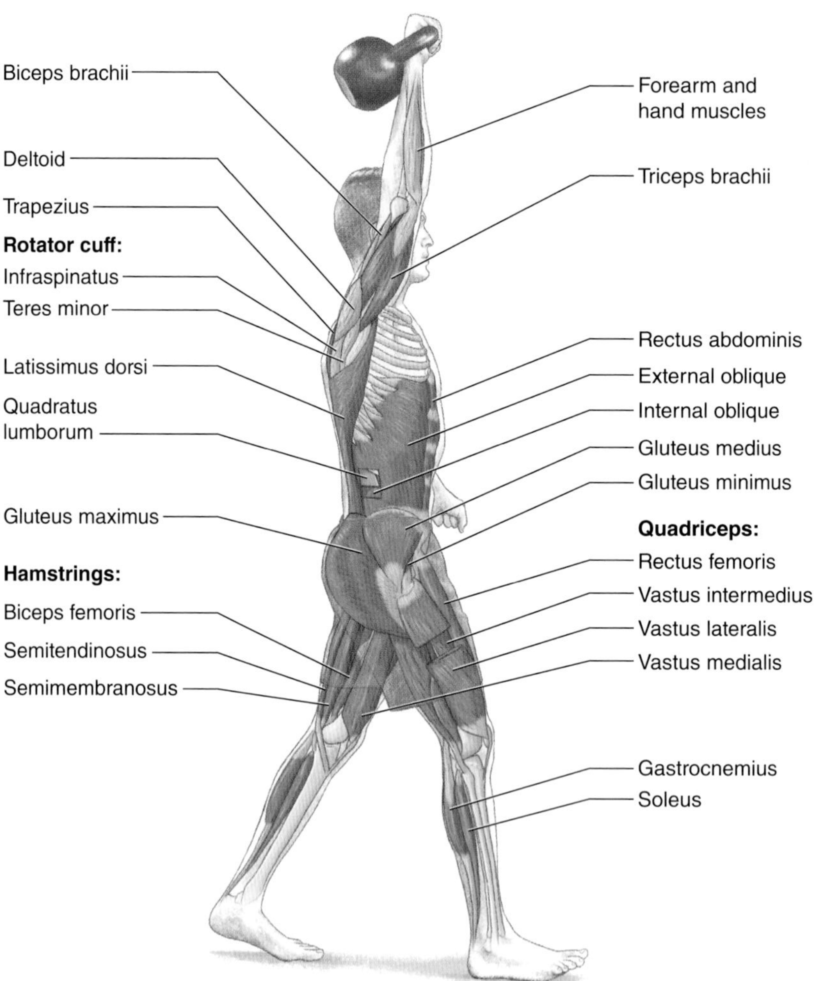

Execution

1. Place your feet about a foot (0.3 m) behind the kettlebell. Take a shoulder-width stance behind the kettlebell with your toes slightly pointed outward.
2. Place the kettlebell overhead, either via a snatch or a clean and then press it overhead.
3. Walk the prescribed distance. Keep your spine as neutral as possible and maintain the vertical plank position while walking. In addition, your shoulders, back, and abdominal muscles should be tight. Your stride length will be shorter than normal while performing the kettlebell overhead carry.

4. When you have reached the prescribed distance, set the kettlebell on the ground in reverse of the motion to stand up with it.
5. Rest, switch hands, and repeat until done.

Muscles Involved

Primary: Latissimus dorsi, trapezius, rhomboids, triceps brachii, rectus abdominis, external and internal obliques, transversus abdominis, gluteus maximus, contralateral quadratus lumborum, gluteus medius, and gluteus minimus, hamstrings (semitendinosus, semimembranosus, biceps femoris), quadriceps (rectus femoris, vastus lateralis, vastus medialis, vastus intermedius), gastrocnemius, soleus

Secondary: Erector spinae (iliocostalis, longissimus, spinalis), deltoid (anterior, middle, and rear), ipsilateral quadratus lumborum, gluteus medius, and gluteus minimus, rotator cuff (infraspinatus, supraspinatus, teres minor, subscapularis), biceps brachii, forearm and hand muscles (wrist flexors, finger flexors)

Anatomic Focus

Hand spacing: Place one hand in contact with the middle of the kettlebell handle, clasping it loosely with your fingers.

Grip: Use a single overhand grip (pronate the hand).

Stance: Position your legs approximately one foot (0.3 m) behind the kettlebell and shoulder-width apart. Point your toes slightly outward. Once the kettlebell is in the overhead position, move your feet in to normal walking width.

Trajectory: Initially, the kettlebell path should be straight up and down and close to the body to get into position. It will travel a horizontal path while walking.

Range of motion: While holding the kettlebell in the overhead position, keep your spine straight and stiff. The erector spinae, abdominal muscles, and latissimus dorsi muscles will help to stabilize and straighten the spine, and the gluteus maximus and hamstrings generate hip extension while moving the kettlebell overhead and then walking with it. Antishrug your shoulders, further activating your latissimus dorsi muscles. Do not overextend your spine. At this point, walk forward for the prescribed distance, then set the kettlebells down and rest.

Holding the kettlebell in the overhead position raises your overall center of mass (which is based on your body weight plus the weight of the kettlebell). Doing this will place more emphasis on keeping your spine neutral and firing your abdominal and gluteal muscles strongly throughout the overhead carry.

(continued)

SINGLE KETTLEBELL OVERHEAD CARRY *(continued)*

Exercise note: The overhead carry allows a lot of variability in sets, repetitions, distance, and time. You can perform it for length of time walking (e.g., three sets of 35 seconds), distance walked (e.g., three sets of 40 yards [37 m]), and as part of your training session or as a warm-up or a finisher. A recommended weight to start with is one fifth to one fourth of your body weight during the overhead carry. Aim for quality over quantity.

Here are a few words of caution: If you are not able to obtain the proper overhead position, do not perform the single kettlebell overhead carry.

VARIATION

Double Kettlebell Overhead Carry

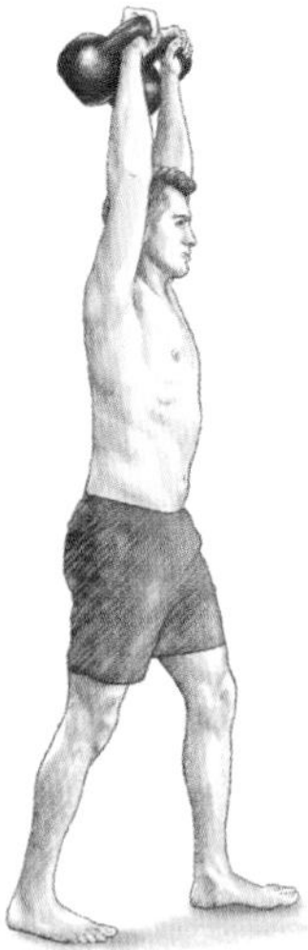

The double kettlebell overhead carry is a bilateral version of the single kettlebell overhead carry. It is performed with two kettlebells. If you are not able to obtain the proper overhead position when performing the double kettlebell overhead carry, do not perform the exercise. As your overall center of mass (which is based on your body weight plus the weight of the kettlebells) rises to its highest position with the weight overhead, this places more importance on your ability to keep your spine neutral by firing your abdominal and gluteal muscles strongly throughout the overhead carry. As with previous exercises in this chapter, there is some variability with how to perform this exercise.

SINGLE KETTLEBELL BOTTOM-UP CARRY

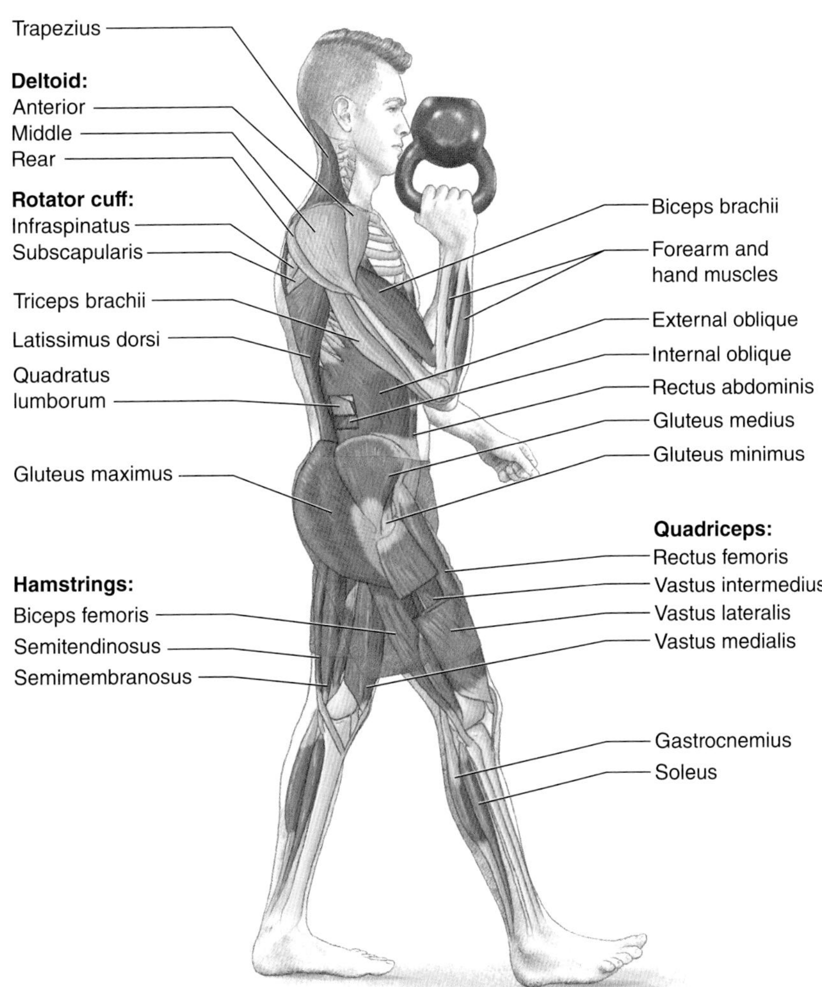

Execution

1. Place your feet about a foot (0.3 m) behind the kettlebell. Take a shoulder-width stance behind the kettlebell with your toes pointed slightly outward.
2. Perform a bottom-up clean with the kettlebell.
3. Walk the prescribed distance. Keep your spine as neutral as possible and maintain the vertical plank position while walking. In addition, your shoulders, back, and abdominal muscles should be tight. Your stride length will be shorter than normal while performing the kettlebell bottom-up carry.

(continued)

SINGLE KETTLEBELL BOTTOM-UP CARRY *(continued)*

4. When you have reached the prescribed distance, set the kettlebell on the ground in reverse of the motion to stand up with it.
5. Rest, switch hands, and repeat until done.

Muscles Involved

Primary: Latissimus dorsi, trapezius, rhomboids, rotator cuff (infraspinatus, supraspinatus, teres minor, subscapularis), rectus abdominis, external and internal obliques, transversus abdominis, gluteus maximus, contralateral quadratus lumborum, gluteus medius, and gluteus minimus, hamstrings (semitendinosus, semimembranosus, biceps femoris), quadriceps (rectus femoris, vastus lateralis, vastus medialis, vastus intermedius), gastrocnemius, soleus, biceps brachii, forearm and hand muscles (wrist flexors, finger flexors)

Secondary: Erector spinae (iliocostalis, longissimus, spinalis), deltoid (anterior, middle, and rear), ipsilateral quadratus lumborum, gluteus medius, and gluteus minimus, triceps brachii

Anatomic Focus

Hand spacing: One hand is in contact with the middle of the kettlebell handle, clasping it loosely with your fingers.

Grip: Use a single overhand grip (pronate the hand).

Stance: Position your legs approximately one foot (0.3 m) behind the kettlebell and shoulder-width apart. Point your toes slightly outward. Once the kettlebell is in the bottom-up position, move your feet in to normal walking width.

Trajectory: Initially, the kettlebell path should be straight up and down and close to the body to get into position. It will travel a horizontal path when you are walking.

Range of motion: While holding the kettlebell in the bottom-up position, keep your spine straight and stiff. Crush grip the handle of the kettlebell to keep it in position. The erector spinae, abdominal muscles, and latissimus dorsi muscles will help to stabilize and straighten the spine, and the gluteus maximus and hamstrings generate hip extension while moving the kettlebell into the bottom-up position and then walking with it. Antishrug your shoulders, further activating your latissimus dorsi muscles. Do not overextend your spine. At this point, walk forward for the prescribed distance, then set the kettlebell down and rest.

Holding the kettlebell in the bottom-up position raises your overall center of mass (which is based on your body weight plus the weight of the kettlebell). Doing this will place more emphasis on keeping your spine neutral and firing your abdominal and gluteal muscles strongly throughout the bottom-up carry.

Exercise note: The bottom-up carry allows a lot of variability in sets, repetitions, distance, and time. You can perform it for length of time walking (e.g., two sets of 25 seconds), distance walked (e.g., two sets of 20 yards [18 m]), and as part of your training session or as a warm-up or a finisher. A recommended weight to start with is one fifth of your body weight during the bottom-up carry. Your goal should be quality over quantity.

VARIATION

Double Kettlebell Bottom-Up Carry

The double kettlebell bottom-up carry is a bilateral version of the single kettlebell bottom-up carry. It is performed with two kettlebells. As your overall center of mass (which is based on your body weight plus the weight of the kettlebells) rises, this places more importance on your ability to keep your spine neutral by firing your abdominal and gluteal muscles strongly throughout the bottom-up carry. As with previous exercises in this chapter, there is some variability with how to perform this exercise.

10

MOBILITY

Range of motion. Mobility and flexibility. Live injury free. Great warm-up for tight joints. These are some of the phrases my kettlebell students have expressed to me over the years regarding mobility training. Preparing yourself for the ensuing training is vitally important to help improve your performance and to decrease chances of injury.

Mobility exercises have received a lot of press over the years. In this chapter we will be discussing kettlebell-specific mobility movements designed to help you get more out of your training session. These exercises are not designed to make you stronger or to build a lot of muscle. They are intended to help you break some rust off your joints and to help reestablish neuromuscular patterns that are needed in your training session. The idea is to move better and prepare yourself for the training ahead.

If you choose to use these exercises in your warm-up, remember to do them to help you in your training session, not to wear you out. After you are done warming up, you should have a light sweat and feel energized to train. A typical warm-up should take 10 to 20 minutes. If it takes longer that 20 minutes, you should rethink your warm-up. More is not usually better.

In my opinion, these are the best kettlebell mobility exercises, but there are more. When choosing a mobility exercise, just be mindful of what it will do for you and how much time it takes to perform. In addition, choose the exercises based on your training session for that day. Unless you have really tight thoracic spine rotation and immovable shoulders, doing an arm bar before kettlebell squatting or swinging is usually not necessary.

When I learned the Cossack squat mobility movement, I was astounded. What it did for me and my hip mobility was amazing. Performing it even helped with rehabbing my ankle after I sprained it badly during a semipro football game. I use a small kettlebell, usually 12 kilograms (26 lb), to help with counterbalance; however, it can be done with no weight as well. Keep your heels down and go as low as you can with your hips.

The kettlebell halo, the pullover, the arm bar, and its cousin, the bent-arm bar, are fantastic for your shoulder complex and your thoracic spine rotation. The latter is sorely lacking in our society today, mostly due to all the electronic devices that we are using. The position of our bodies while using our phones and other electronics causes increased thoracic kyphosis (curvature in a sagittal plane) and protracted shoulders. Using these movements will slowly peel back the layers and allow you to do things with this part of your body that you haven't been able to do for years.

One of the things you will discover from performing the arm bar is when you start, your upper back muscles will be sore for a few days as they are getting used to working again. When I played football, the arm bar was a staple of my training session warm-ups as it helped to reset my shoulders after playing defensive tackle for three or more hours a few days earlier.

The prying goblet squat (chapter 6) is also one of my favorites for what it does for your lower extremities and your hip region. This is definitely a mobility move and not a strength move. I had a kettlebell student who used to do a lot of slalom water skiing, always having the same foot forward. When I showed her how to do this mobility move, she went to one side with no issues, but when she went to the other side, she made it to the middle and couldn't go any farther. She worked on improving this range of motion and consequently her squat and swing technique improved dramatically.

Another cue that I share with my students while they are performing the prying goblet squat is to pull the two halves of your pelvis as far apart as possible from each other. Doing this makes space with your hip or pelvic complex, allowing you to squat deeper and keep your spine neutral. I also recommend using a lighter kettlebell when doing this movement.

Mobility movements are analogous to the backstage people that you see in the credits after watching a movie. They may not be the big actors in the movie, but without them, there would be no show to watch. Start your training session off on the right foot by doing a short yet concise warm-up and then proceed to the main exercises that you are training that day. Remember this mantra when performing the mobility exercises: increase performance, decrease injury.

EXERCISE FOCUS FOR MOBILITY EXERCISES

The mobility exercises in this chapter can be organized into two categories:

1. Lower body
2. Upper body

Each category of mobility exercises provides different ways to help you learn proper technique and develop mobility. Here, we take a closer look at the focus of the exercises in each category.

When performing mobility exercises, the primary focus should be to get your body ready for the main exercises that will follow. Mobility exercises are not strength exercises and should not be treated as such. When performing them for the first couple sessions, you may experience some mild muscle soreness as you are still going through eccentric and concentric contractions while performing them. However, the soreness should dissipate within a few days.

Lower Body

Two important mobility movements for the lower body are the Cossack squat, described later in this chapter, and the prying goblet squat, covered in chapter 6. Each has its own purpose, yet I strongly recommend that you perform both when doing lower body training following your mobility warm-up. Some of the benefits of performing the Cossack squat is not only mobilizing your lower body, namely your posterior chain, but it is also enhancing your strength movements later in your training session. It also helps to promote neuromuscular control of your ankle, knee, and hip areas, which are necessary for daily activities and sports such as martial arts, strongman, gymnastics, and soccer (figure 10.1).

FIGURE 10.1 Soccer requires precise neuromuscular control of the ankle, knee, and hip, a characteristic developed with the help of the Cossack squat.

Keeping your knees tracking the toes will also benefit you during the squat and other kettlebell movements. During the prying goblet squat, described in chapter 6, in addition to the above benefits, you are making space in your lumbopelvic region to be able to do many different kettlebell movements safely that will enhance your performance and help decrease injury.

Upper Body

For the upper body, the kettlebell halo, armbar, and pullover are essential to not only help improve your torso mobility and shoulder mobility but also to prime those areas for the primary movements that will follow your mobility warmup. All three of those movements are similar in their positive effects for the person performing them, but they also have their individual benefits. Athletes in swimming, track and field, martial arts, and volleyball (figure 10.2) are some of the people who will benefit from performing these mobility movements.

FIGURE 10.2 In volleyball, both setters and spikers must have good mobility in the upper body.

The kettlebell halo will help increase your torso stability as you move the kettlebell around your head and neck. The arm bar, and its cousin the bent arm bar, help improve your shoulder and thoracic spine mobility and torso rotary capacity while at the same time developing strong, mobile shoulders and upper back muscles. The kettlebell pullover not only helps you to develop strong mobile shoulders and upper back muscles but also to develop a stronger core by keeping your back flat and your abdominal muscles braced while moving the kettlebell over your head. All three of these movements will also assist in improving the neuromuscular control of your shoulders, which is paramount in helping you prevent injury and increase your performance.

COSSACK SQUAT

Execution

1. You have two options to get the kettlebell into position to perform the Cossack squat:
   a. Place your feet about a foot (0.3 m) behind the kettlebell. Take a shoulder-width stance behind the kettlebell with your toes pointed slightly outward. Hike the kettlebell back with both hands on the handle (as in the swing discussed in chapter 3) and stand up, keeping the kettlebell close to your body as you bend both elbows. You will end up holding the kettlebell with both hands, one on each side of the handle.
   b. Stand over the kettlebell with a shoulder-width stance and your toes pointed slightly outward (as in the deadlift exercises in chapter 2). Grip the handle with both hands. As you stand up to deadlift the kettlebell, continue the motion and bend your elbows so your hands end up being on each side of the handle at your chest level.

(continued)

COSSACK SQUAT *(continued)*

2. Move your feet to more than shoulder-width apart and slightly point your feet out. If needed, you can adjust your stance width in the bottom of the movement.
3. Laterally transfer your weight toward your right foot while keeping your left foot on the ground. Keep your right knee tracking your toes while descending your hips below your knees, if possible. Make sure that your right foot is entirely planted and your left leg is straight. Keep your chest high while holding the kettlebell. An alternative is to allow your left toes to come off the ground, pointing toward the ceiling. Descend as far as you are comfortable and hold for few seconds, breathing at the same time. When done, come back up, moving toward the other leg.
4. Shift your weight toward your left leg either by standing up and achieving the vertical plank position before descending for the next repetition or by coming up just enough to allow your body to move to the other side. Make sure that your hips and shoulders ascend at the same time.
5. If you stand up fully after each repetition, assume the vertical plank position before starting the next repetition and repeat until you reach the desired number of repetitions.

Muscles Involved

Primary: Erector spinae (iliocostalis, longissimus, spinalis), quadriceps (rectus femoris, vastus lateralis, vastus medialis, vastus intermedius), gluteus maximus, gluteus medius, and gluteus minimus, abductors and adductors, hamstrings (semitendinosus, semimembranosus, biceps femoris), hip flexors (iliopsoas, rectus femoris, sartorius), rectus abdominis, transversus abdominis, internal oblique, external oblique

Secondary: Trapezius, rhomboids, latissimus dorsi, quadratus lumborum, lower leg muscles (tibialis anterior and posterior, peroneus longus and brevis), calf muscles (gastrocnemius, soleus), forearms (wrist flexors, finger flexors), elbow flexors (biceps brachii, brachioradialis)

Anatomic Focus

Hand spacing: Place your hands on opposite sides of the handle of the kettlebell.

Grip: Use a double overhand grip (supinate your hands at the top).

Stance: Position your legs more than shoulder-width apart. Point your toes slightly outward.

Trajectory: Move the kettlebell in the vertical plane from the top to the bottom.

Range of motion: At the beginning of each repetition, your toes will be rooted to the ground, your kneecaps pulled up, your gluteal muscles squeezed, and your abdominal and latissimus dorsi muscles strongly engaged. As in the kettlebell deadlift, the erector spinae, abdominal muscles, and latissimus dorsi muscles will help to stabilize and straighten the spine while the quadriceps, abductors and adductors of the hip, and the gluteal muscles and hamstrings generate the squatting motion. Antishrug your shoulders, further activating your latissimus dorsi muscles. Your spine should be straight and stiff throughout the movement, even when your hips descend below your knees.

Being that you are using a single kettlebell during this entire movement and shifting your weight from side to side, the load on the body will be asymmetrical. When you keep the straight-leg foot on the ground throughout the movement, you will tend to feel it more in your adductors. When the toes are off the ground and pointing up, you will feel it more in the hamstrings of the straight leg.

Technique note: Do not use heavy loads when performing the Cossack squat, as this is a mobility movement. One of its many benefits is also increasing your neuromuscular control of your lower limbs, which can help decrease or prevent injury in athletics or general life.

KETTLEBELL HALO

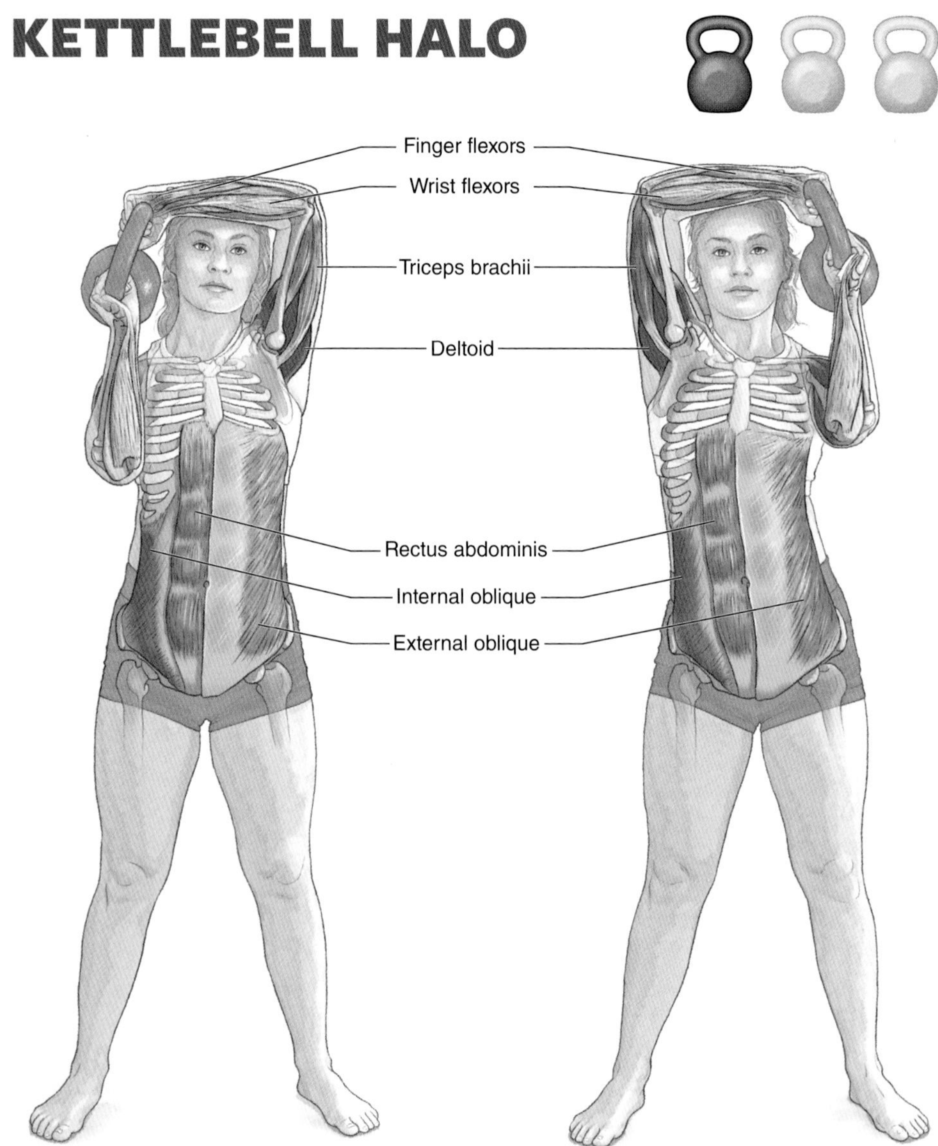

Execution

1. Grab a light kettlebell and hold it with both hands on either side of the handle, upside down, while standing. The bottom of the kettlebell is facing up.
2. Root your feet to the ground, pull your kneecaps up, squeeze your gluteal muscles, and brace your abdominal muscles. Keep your ribs stacked on top of your pelvis and your wrists neutral throughout this movement.
3. Slowly move the kettlebell in a circle around you. Start by moving it over your left shoulder, allowing the bottom to face behind you. Once it is over your shoulder, keep the kettlebell as close to your torso as

possible. At this point the bottom will face the floor. As you continue to move it around your head and neck, move it back over your right shoulder, again having the bottom face behind you. Finish back at the start position, in front of your chest. Pause briefly and continue.

4. Think of your head and neck area as the sun and the kettlebell as the earth, rotating around it.
5. Move in this direction for a set number of repetitions, then repeat it in the opposite direction for the same number of repetitions. At first, you may not be able to keep the kettlebell close to your torso, but as the repetitions continue, you will get closer and have more mobility.
6. When you are done, set the kettlebell down on the ground.

Muscles Involved

Primary: Deltoid (anterior, medial, and posterior), trapezius, rhomboids, rectus abdominis, transversus abdominis, internal oblique, external oblique, triceps brachii

Secondary: Gluteus maximus, forearms (wrist flexors, finger flexors)

Anatomic Focus

Hand spacing: Place your hands on opposite sides of the handle of the kettlebell.

Grip: Use a double overhand grip (place your hands in a neutral position on the handle).

Stance: Position your feet will be hip-width apart.

Trajectory: The kettlebell will move in the transverse plane, around your head and neck, in a clockwise fashion, then counterclockwise, or vice versa.

Range of motion: Throughout the halo movement, your toes will be rooted to the ground, your kneecaps pulled up, your gluteal muscles squeezed, and your abdominal muscles strongly engaged. Your spine should be straight and stiff throughout the movement. Keep your ribs stacked on top of your pelvis and your wrists neutral throughout this movement. Move the kettlebell around your head and neck, not the other way around. Using a mirror while doing the halo movement is encouraged as you get visual feedback as to how you are performing the exercise. You may switch the rotational direction of the kettlebell with each repetition, instead of performing all repetitions on one side then all repetitions on the other side. Usually two or three sets of five repetitions per side is recommended.

Technique note: Do not use heavy loads when performing the kettlebell halo, because this is a mobility movement. One of its many benefits is increasing your neuromuscular control of your upper body and limbs, which can help decrease or prevent injury in athletics or general life.

(continued)

KETTLEBELL HALO *(continued)*

VARIATION

Tall Kneeling Kettlebell Halo

To progress the kettlebell halo, kneel on the ground with your knees hip-width or slightly wider, your lower legs parallel to each other, and your toes flat on the ground. Squeeze your gluteal muscles, brace your abdominal muscles, and perform the halo movement.

Half Kneeling Kettlebell Halo

For an advanced progression of the kettlebell halo, get into a kneeling lunge position with one foot forward and the opposite knee on the ground. Make sure that your knee, hips, shoulders, and head are stacked on top of each other and are squared up. Place little weight on the front foot. Squeeze your gluteal muscles, brace your abdominal muscles, and perform the halo movement. When this becomes easier, I recommend narrowing your stance in the lunge position by moving the front foot in line with the rear leg.

KETTLEBELL ARM BAR

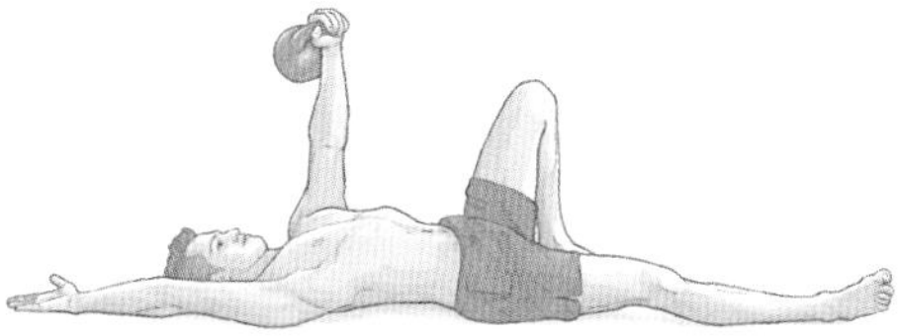

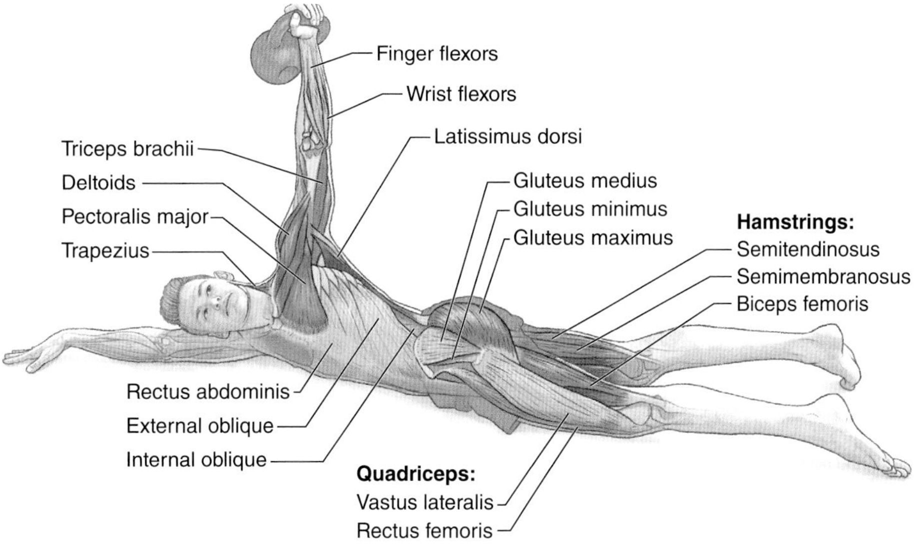

Execution

1. Lie on your left side with the kettlebell close to you. Grab the handle with your left hand and place your right hand with a thumbless grip over the top. While keeping the kettlebell close to you, roll to your back with the two-handed grip. The kettlebell will be down by your ribs with your left elbow on the ground.
2. Let go with your right hand and floor press the kettlebell up or press the kettlebell up with both hands.
3. Keep your thumb pointed toward your head. Bend your left knee to approximately 90 degrees and keep your right leg straight. Place your right hand on the ground above your head.
4. With your left leg, roll your body to the right, using your right arm and leg as the axis. Place your left knee on the ground with your hip flexed to 90 degrees while resting your head on your right arm.

(continued)

KETTLEBELL ARM BAR *(continued)*

5. Keep your left arm perpendicular to the floor and your thumb pointing toward your head. Antishrug both shoulders and squeeze your shoulder blades together, packing your shoulders during this movement.
6. Straighten out your left leg so that your legs are approximately shoulder-width or more apart. Keep your toes pointed.
7. Breathe in, and as you exhale, squeeze your left gluteus maximus to bring the front of your left hip toward the ground. Do this for three to five repetitions. When you are done, flex your left hip back to 90 degrees. Using the fingers of your right hand, slowly inch your right arm into further flexion. Keep your head on your right biceps. If you reach full flexion of your right arm, do not go any farther.
8. Straighten the leg from 90 degrees of hip flexion to 0 degrees of hip flexion. Set the kettlebell on the ground by reversing the movements and positions that you used to perform the kettlebell arm bar.
9. When switching sides, you can use both hands to drag the kettlebell on the floor around your head in an arc to the other side, or you can move your body to the other side of the kettlebell.

Muscles Involved

Primary: Erector spinae (iliocostalis, longissimus, spinalis), gluteus maximus, hamstrings (semitendinosus, semimembranosus, biceps femoris), latissimus dorsi, pectoralis major and minor, deltoids (anterior, medial, and posterior), rhomboids, trapezius

Secondary: Gluteus medius and gluteus minimus, quadriceps (rectus femoris, vastus lateralis, vastus medialis, vastus intermedius), rectus abdominis, transversus abdominis, internal oblique, external oblique, forearms (wrist flexors, finger flexors), elbow extensors (triceps brachii)

Anatomic Focus

Grip: Position the kettlebell handle in a diagonal position from the web of the thumb to the pisiform of the hand. Your wrist should be neutral.

Stance: Place your feet shoulder-width or slightly wider apart.

Trajectory: During the arm bar, the kettlebell will be essentially motionless.

Range of motion: The kettlebell arm bar is a fantastic mobility movement for your thoracic spine and shoulder. Pause at each section while moving into position. Pay attention to what your body is saying to you. If you need to move your foot or your hand before performing the next move, do so. Check that you are in the best position to succeed before performing it.

When done properly, it will help bulletproof your shoulders and stretch your pectoralis and latissimus dorsi muscles. Keep your head resting on the arm on the floor. This will help decrease tension in your loaded arm.

Technique note: This movement is a mobility movement, not a strength movement. Therefore, use a weight that is easy to moderate but not heavy. Do it as one of your warm-up exercises prior to your training session.

VARIATION

Kettlebell Bent-Arm Bar

This variation is set up like the arm bar, except that you stay on your side with your top hip flexed to 90 degrees and with the knee on the ground. With your elbow extended, thumb pointed toward your head, wrist neutral, and forearm perpendicular to the floor, flex and pull your elbow down and behind your thoracic spine, if possible. Keep your head on the arm on the floor. Think of this as though you are scratching your back with your elbow. Hold the position for a brief time, then press the kettlebell back up. Perform three to five repetitions, set the kettlebell on the ground safely, and switch sides.

KETTLEBELL PULLOVER

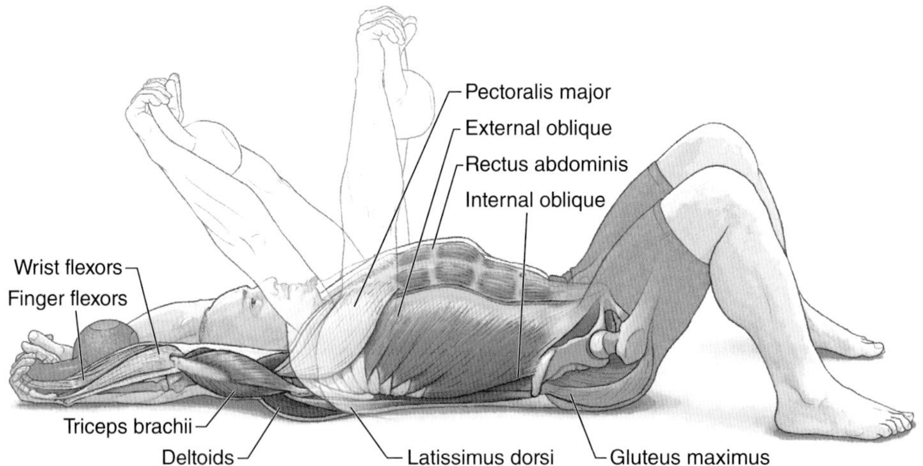

Execution

1. Lie on your back with your knees bent and your feet flat on the floor.
2. Safely lift a kettlebell above your chest with your triceps locked and elbows straight. Hook the handle with your thumbs; pronate your hands.
3. Flatten your lower back to the floor and brace your abdominal muscles.
4. Slowly move the kettlebell above your head and go as far as you can, stopping when you reach your limit or at the floor.
5. Slowly return the kettlebell to the start position, keeping your lower back flat to the floor and your abdominal muscles braced.
6. Repeat. When you have done three to five repetitions, set the kettlebell safely back on the ground.

Muscles Involved

Primary: Erector spinae (iliocostalis, longissimus, spinalis), latissimus dorsi, anterior and medial deltoid, rhomboid, trapezius, triceps brachii, rectus abdominis, transversus abdominis, internal oblique, external oblique

Secondary: Gluteus maximus, pectoralis major and minor, forearms (wrist flexors, finger flexors)

Anatomic Focus

Hand spacing: Place your hands next to each other on the kettlebell handle.

Grip: Hook the kettlebell handle by your thumbs with your hands pronated.

Stance: Place your feet hip-width apart or narrower.

Trajectory: During the pullover, the kettlebell will move in a semicircle toward the ground above your head.

Range of motion: The kettlebell pullover is a fantastic mobility movement for your shoulders, namely your latissimus dorsi and your glenohumeral and acromioclavicular joints. Briefly pause at the top and the bottom while performing this movement. Pay attention to what your body is saying to you. When done properly, this exercise will help make your shoulders bulletproof and stretch your pectoralis and latissimus dorsi muscles.

Technique note: This movement is a mobility movement, not a strength movement. Therefore, use a weight that is easy to manipulate but not heavy. Do it as one of your warm-up exercises prior to your training session.

VARIATION

Straight Legs

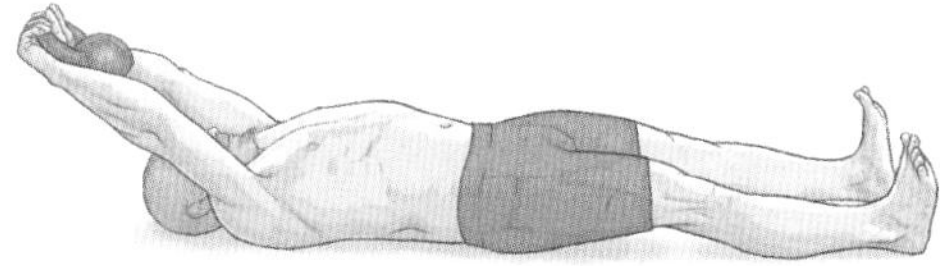

An advanced progression of the kettlebell pullover is to keep your legs straight. Lock your knees and dorsiflex your ankles. Keep your lower back flattened to the floor and your abdominal muscles braced throughout this movement.

Single-Arm Kettlebell Pullover

This variation is set up like the kettlebell pullover, except you use only one hand. Choose a lighter kettlebell than you used with the two-handed version. Hook the handle as before and slowly move the kettlebell above your head. You may find differences in your mobility per side. Hold for a brief time, then slowly return the kettlebell to the start position. Perform three to five repetitions per side, set the kettlebell on the ground safely, and switch sides.

EXERCISE FINDER

DEADLIFT

Single Kettlebell Deadlift, Inside Legs 16
Double Kettlebell Deadlift, Inside Legs 18
Single Kettlebell Deadlift With One Hand, Inside Legs 20
Double Kettlebell Deadlift, Outside Legs 22
Single Kettlebell Deadlift With One Hand, Outside Legs (Suitcase Deadlift) 24
Single-Leg Deadlift With Body Weight 26
Kickstand Deadlift 28
Single-Leg Deadlift With Eyes Closed 29
Single-Leg Deadlift With Single Kettlebell 30
Floor Single-Leg Deadlift With Single Kettlebell 31
Single-Leg Deadlift With Double Kettlebells 32
Floor Single-Leg Deadlift With Double Kettlebells 33

SWING

Single Kettlebell Two-Hand Swing, Inside Legs 40
Dead Stop Swing (Power Swing) 42
Single Kettlebell One-Hand Swing, Inside Legs 43
Single Kettlebell One-Hand Swing, Alternating Hands, Inside Legs 46
Double Kettlebell Swing, Inside Legs 49
Single Kettlebell One-Hand Swing, Outside Legs 52
Double Kettlebell Swing, Outside Legs 55

CLEAN AND PRESS

Single Kettlebell Clean 64
Dead Stop Clean 66
Double Kettlebell Clean 67
Single Kettlebell Press 70
Double Kettlebell Press 72
Alternating Press 74
See-Saw Press 74
Sots Press With Double Kettlebells 75
Alternating Sots Press 77
See-Saw Sots Press 77
Single Kettlebell Floor Press 78
Straight Legs 80
Bottom-Up Press 80
Double Kettlebell Floor Press 81
Straight Legs 83
Alternating Press 83
Double Kettlebell Bridge Floor Press 84
Alternating Press 86
Single Kettlebell Bottom-Up Clean 87
Single Kettlebell Bottom-Up Press 90
Double Kettlebell Bottom-Up Clean 91
Double Kettlebell Bottom-Up Press 94

GET-UP

Body-Weight (i.e., "Naked") Get-Up 98
Shoe Get-Up 102
Sandbag Get-Up 105
Kettlebell Get-Up 106
Bottom-Up Get-Up 109
Kettlebell Get-Up Plus Press 110

SQUAT

Goblet Squat 119
Prying Goblet Squat 122
Single Kettlebell Front Squat 125
Double Kettlebell Front Squat 127
Cossack Squat 129
Hack Squat 132

SNATCH

Single Kettlebell T. Rex Swing With One Hand 139
Single Kettlebell Snatch With One Hand 142
Double Kettlebell Snatch 145
Single Kettlebell Snatch to Eccentric Press 148
Double Kettlebell Snatch to Eccentric Press 151

ROW AND PULL-UP

Renegade Row 156
Standing Single Kettlebell Row 159
Standing Double Kettlebell Row 161
Pull-Up With a Kettlebell 162

CARRY

Double Kettlebell Farmer's Carry 168
Single Kettlebell Farmer's Carry (Suitcase Carry) 170
Double Kettlebell Rack Carry 171
Single Kettlebell Rack Carry 173
Single Kettlebell Overhead Carry 174
Double Kettlebell Overhead Carry 176
Single Kettlebell Bottom-Up Carry 177
Double Kettlebell Bottom-Up Carry 179

MOBILITY

Cossack Squat 185
Kettlebell Halo 188
Tall Kneeling Kettlebell Halo 190
Half Kneeling Kettlebell Halo 190
Kettlebell Arm Bar 191
Kettlebell Bent-Arm Bar 193
Kettlebell Pullover 194
Straight Legs 195
Single-Arm Kettlebell Pullover 195

ABOUT THE AUTHOR

Dr. Michael Hartle is a full-time chiropractic physician. He is also certified as a nutritionist (DACBN), chiropractic sports physician (CCSP), and strength and conditioning specialist (CSCS). He is also a StrongFirst Certified Master Instructor with StrongFirst (StrongFirst.com). He is the co-developer of the StrongFirst SFL barbell certification. Dr. Hartle travels around the United States and the world teaching barbell and kettlebell certification courses. He is also currently working toward his PhD in exercise science at Concordia University Chicago. He has been practicing in Fort Wayne, Indiana, for the last 27 years.

A former nationally ranked powerlifter who has won several national titles with USA Powerlifting (USAPL), Dr. Hartle was the vice president for two years in addition to being the drug testing chairperson for five years. He was also the chairperson of their Sports Medicine Committee for over 20 years, a committee he created in 1994. Starting in 1998, he was the head coach of the USAPL world bench press team for eight years, coaching the U.S. team to the 2004 International Powerlifting Federation World Championship team title. His best competition lifts are 705-pound squat, 535-pound bench press, and 635-pound deadlift, with a best combined total of the three lifts of 1,840 pounds in the former 275-pound weight class. For the next 10 years, he was playing semi-pro football as a defensive tackle—and loving it! His football team, the Adams County Patriots, won the National AA Semi-Pro Football Championship in 2008 and were undefeated for two years straight.

He treats patients from babies to senior citizens, and he trains and advises athletes of all levels, from youth athletes to NCAA student-athletes and professional athletes. He coached junior high football and track and field, volunteering his time, for 16 years. He has three sons and three grandchildren who keep him busy with their personal endeavors, including crawling, hockey, football, lacrosse, track and field, and, of course, academics.

Find other outstanding resources at
US.HumanKinetics.com
Canada.HumanKinetics.com
In the U.S. call 1-800-747-4457
Canada 1-800-465-7301
International 1-217-351-5076
HUMAN KINETICS

ANATOMY SERIES

Each book in the *Anatomy Series* provides detailed, full-color anatomical illustrations of the muscles in action and step-by-step instructions that detail perfect technique and form for each pose, exercise, movement, stretch, and stroke.

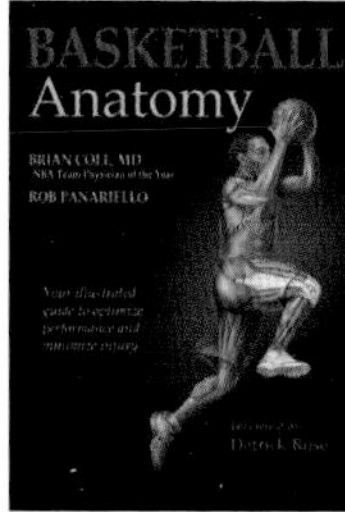

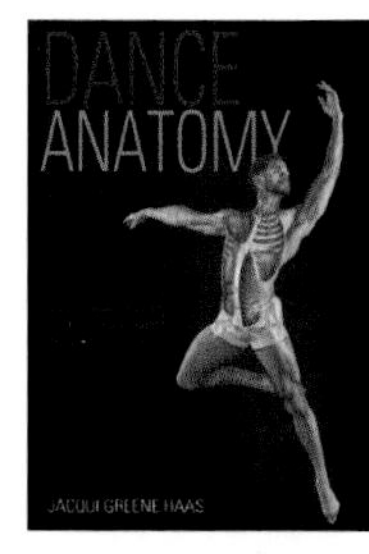

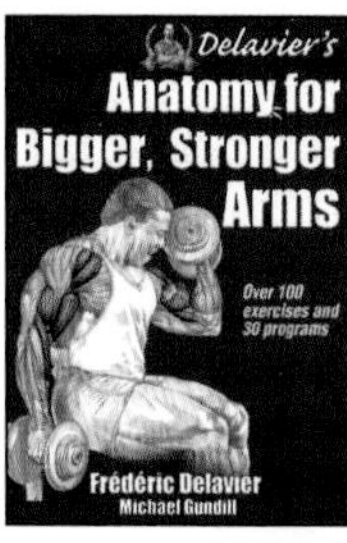

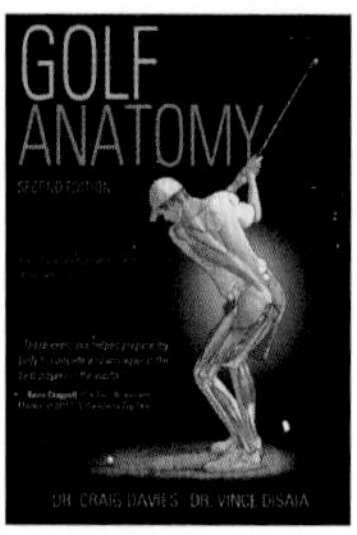

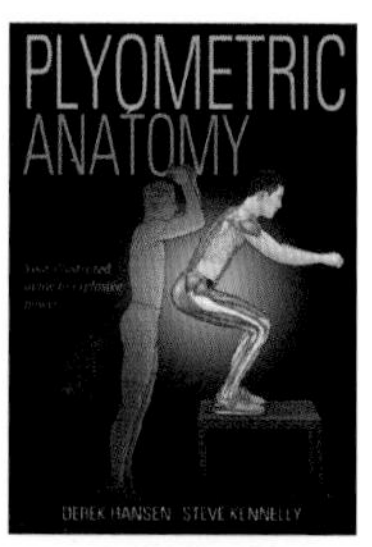

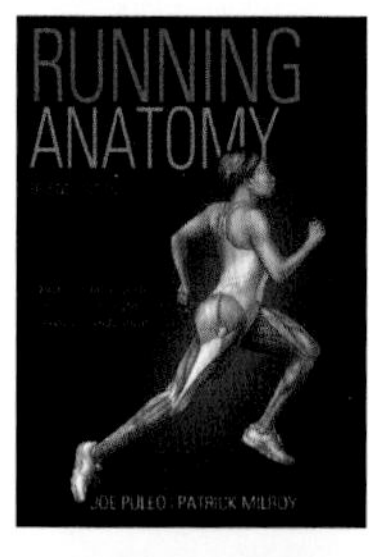

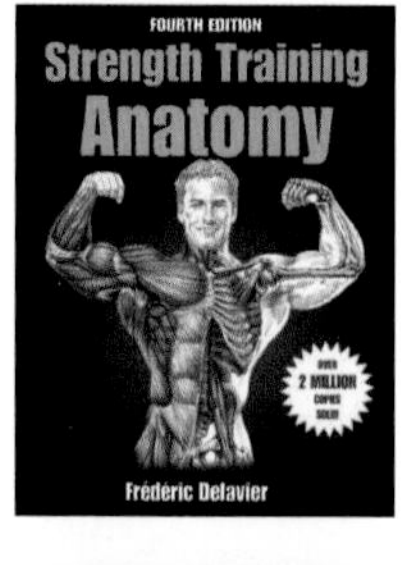

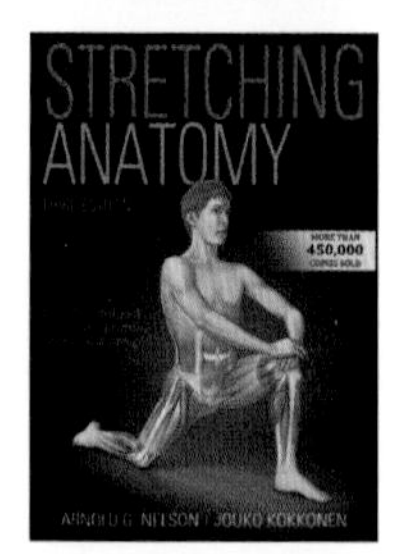

U.S. 1-800-747-4457 • US.HumanKinetics.com/collections/anatomy
Canada 1-800-465-7301 • Canada.HumanKinetics.com/collections/anatomy
International 1-217-351-5076